AP Calculus AB Lecture Notes

By Rita Korsunsky

Copyright © 2019 by Rita Korsunsky

All rights reserved.

ISBN: 1500763845
ISBN-13: 978-1500763848
No part of this book may be reproduced or transmitted in any form or by any means, including electronic transmission and photocopying.

PREFACE

Imagine having AP Calculus AB Lecture Notes that illustrate every problem, walking you through the procedure step-by-step. Imagine having every proof, illustration, or theorem explained concisely and accurately.

Well, with this book, AP Calculus AB Lectures Notes, you can!

Why is this paperback so convenient?

This book contains printouts of all the Powerpoint presentations on topics covered by both the AP Calculus AB Exam and the first part of the BC Exam. You can take notes on this book, study from it, and use it as test preparation material for chapter tests as well as for the AP test. At the end of this book, you will find the list of all the formulas and theorems needed for the AP test. These lecture notes can be used for both review and learning, and are a perfect fit for every student no matter their current knowledge of Calculus. Every example and every lesson targets a specific skill or formula. With this book, you will have every concept you need to know at the tip of your fingers.

This book reflects the current changes in the College Board requirements.

Our books are written by Mrs. Rita Korsunsky, a High School Mathematics Teacher with more than twenty years of experience teaching AP Calculus BC. Her lectures are rigorous, entertaining, and effective. Her students' AP Scores speak for themselves:

Around 100% of her students pass the AP Exam

Around 94% of her students get 5 on the AP Exam

For more information and testimonials please visit www.mathboat.com

A note from the Author:

I would like to thank all of my past, present, and future students for inspiring me to publish this Book. A special thanks to Lillian Li for hand drawing the exquisite illustrations, Travis Chen for assisting me in converting the Powerpoints and my sons Boris and David for their technical assistance. A special debt of gratitude is due to my husband Alex for his continual support and help every step of the way.

Please send all questions and concerns to:

captain@mathboat.com

Sincerely,

Rita Korsunsky

www.mathboat.com

TABLE OF CONTENTS

Chapter 1 Limits of functions………………… page 1

Chapter 2 The Derivative…………………… page 13

Chapter 3 Applications of Derivative………… page 38

Chapter 4 Integrals…………………………… page 56

Chapter 5 Applications of Definite Integrals…… page 84

Chapter 6 Logarithmic & Exponential Functions… page 97

Chapter 7 Inverse Trigonometric Functions…… page 111

Formulas and Theorems…………………………... page 116

www.mathboat.com

1.1 Introduction to Limits

Limit of a function

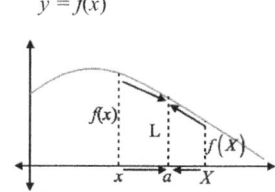

When x is approaching to a, value of $f(x)$ is getting closer and closer to L and becomes almost equal to L

Notation:
$$\lim_{x \to a} f(x) = L$$

We say, that the limit of $f(x)$ as x approaches to a, is L

Intuitive Meaning: We can make $f(x)$ as close to L as desired by choosing x sufficiently close to a, and $x \neq a$

Example 1

Find $\lim_{x \to 3}(x+2)$

x	f(x)=x+2
2.9	4.9
2.99	4.99
2.999	4.999

x	f(x)=x+2
3.1	5.1
3.01	5.01
3.001	5.001

$\lim_{x \to 3}(x+2) = \boxed{5}$

The closer x is to 3, the closer $(x+2)$ is to 5

Example 2

Find $\lim_{x \to 3} g(x)$ if $g(x) = \dfrac{x^2 - x - 6}{x - 3}$

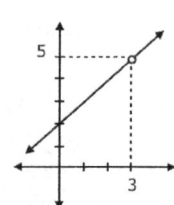

Notice that $\dfrac{x^2 - x - 6}{x - 3} = \dfrac{(x+2)\cancel{(x-3)}}{\cancel{(x-3)}}$

$= x + 2$

There is a hole at $x = 3$

From Example 1:
$\lim_{x \to 3} \dfrac{x^2 - x - 6}{x - 3} = \lim_{x \to 3}(x+2) = 5$

Even though $g(3) \neq 5$, the limit is still 5

Limit of a function

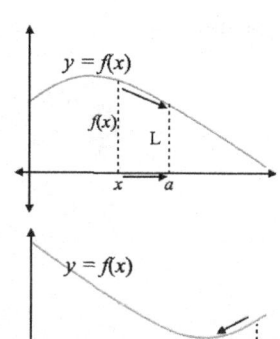

Notation: $\lim_{x \to a^-} f(x) = L$ (left-hand limit)

We say, that the limit of $f(x)$ as x approaches to a from the left, is L

Intuitive Meaning: We can make $f(x)$ as close to L as desired by choosing x sufficiently close to a, and $x < a$.

Notation: $\lim_{x \to a^+} f(x) = L$ (right-hand limit)

We say, that the limit of $f(x)$ as x approaches to a from the right, is L

Intuitive Meaning: We can make $f(x)$ as close to L as desired by choosing x sufficiently close to a, and $x > a$.

Theorem

$\lim_{x \to a} f(x) = L$ if and only if
$\lim_{x \to a^+} f(x) = L = \lim_{x \to a^-} f(x)$

The limit of $f(x)$ as x approaches a exists if and only if both the right-hand and left-hand limits exist and are equal.

Example 3

Using the graph of $f(x) = \dfrac{1}{x}$, explain why $\lim\limits_{x \to 0} \dfrac{1}{x}$ does not exist.

$f(x) = \dfrac{1}{x}$

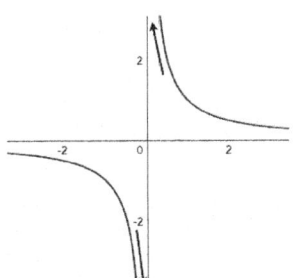

As the graph shows, f(x) does not approach a number *L* as *x* approaches 0 from the left and from the right, so **limit does not exist.**

Example 4

Sketch the graph of $f(x) = \sqrt{x-3}$ and find, if possible,

a) $\lim\limits_{x \to 3^+} f(x)$ b) $\lim\limits_{x \to 3^-} f(x)$ c) $\lim\limits_{x \to 3} f(x)$

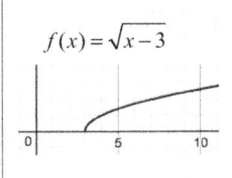

a) $\lim\limits_{x \to 3^+} \sqrt{x-3} \approx \sqrt{3.00001 - 3} \approx \boxed{0}$

b) $\lim\limits_{x \to 3^-} \sqrt{x-3} \approx \underbrace{\sqrt{2.99999 - 3}}_{<0} = \boxed{D.N.E}$

c) $\lim\limits_{x \to 3^+} f(x) \ne \lim\limits_{x \to 3^-} f(x) \Rightarrow$

$\boxed{\lim\limits_{x \to 3} f(x) \text{ D.N.E.}}$

Example 5

Sketch the graph of $f(x) = \dfrac{|x|}{x}$ and find, if possible

a) $\lim\limits_{x \to 0^-} f(x)$ b) $\lim\limits_{x \to 0^+} f(x)$ c) $\lim\limits_{x \to 0} f(x)$

$f(x) = \dfrac{|x|}{x}$

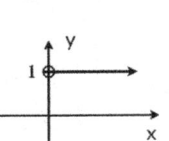

a) $\lim\limits_{x \to 0^-} f(x) = \boxed{-1}$

b) $\lim\limits_{x \to 0^+} f(x) = \boxed{1}$

c) $\lim\limits_{x \to 0} f(x) = \boxed{D.N.E}$

(left-hand and right-hand limits are not equal)

Example 6

Use the graph of $f(x)$ to find

$\lim\limits_{x \to 1} f(x), \lim\limits_{x \to 2} f(x), \lim\limits_{x \to 3} f(x)$.

$\lim\limits_{x \to 1^-} f(x) = 0$

$\lim\limits_{x \to 1^+} f(x) = 1$

$\lim\limits_{x \to 1} f(x)$ DNE   Even though f(1) = 1

$\lim\limits_{x \to 2^-} f(x) = 1$

$\lim\limits_{x \to 2^+} f(x) = 1$

$\lim\limits_{x \to 2} f(x) = 1$ Even though f(2) = 2

$\lim\limits_{x \to 3} f(x) = 2$

$\lim\limits_{x \to 3^+} f(x) = 2$

$\lim\limits_{x \to 3} f(x) = 2$

Example 7

Let *f* be defined as follows,

$f(x) = \begin{cases} x^2 + 1, & \text{for } -2 \le x < 2 \\ 1, & \text{for } x = 2 \\ 5, & \text{for } 2 < x < \infty \end{cases}$

What is the limit of $f(x)$ as *x* approaches 2?

$\lim\limits_{x \to 2^-} f(x) = \lim\limits_{x \to 2^-} (x^2 + 1) = 2^2 + 1 = 5$

$\lim\limits_{x \to 2^+} f(x) = 5$

$\lim\limits_{x \to 2^-} f(x) = \lim\limits_{x \to 2^+} f(x) = 5$ $\Rightarrow \boxed{\lim\limits_{x \to 2} f(x) = 5}$

Example 8

Find the limit of $f(x)$ as *x* approaches 1.

$f(x) = \begin{cases} -x^3 & \text{If } x < 1 \\ 3 & \text{If } x = 1 \\ x - 3 & \text{If } x > 1 \end{cases}$

$\lim\limits_{x \to 1^-} f(x) = \lim\limits_{x \to 1^-} (-x^3) = -1$

$\lim\limits_{x \to 1^+} f(x) = \lim\limits_{x \to 1^+} (x - 3) = -2$

$\lim\limits_{x \to 1^-} f(x) \ne \lim\limits_{x \to 1^+} f(x)$ $\boxed{\lim\limits_{x \to 1} f(x) \text{ DNE}}$

1.2 Techniques for Finding Limits

 Mathboat.com

Theorem

If c and a are constants and x is a variable, then

The limit of a constant is the constant:

$$\lim_{x \to a} c = c$$

For Example:

$$\lim_{x \to 2} 6 = 6$$

and $\lim_{x \to a} x = a$

$$\lim_{x \to \sqrt{3}} x = \sqrt{3}$$

Properties of Limit

If $\lim_{x \to c} f(x)$, $\lim_{x \to c} g(x)$, a and c are real numbers, then:

1) $\lim_{x \to c}[f(x) + g(x)] = \lim_{x \to c} f(x) + \lim_{x \to c} g(x)$

2) $\lim_{x \to c}[f(x) - g(x)] = \lim_{x \to c} f(x) - \lim_{x \to c} g(x)$

3) $\lim_{x \to c}[f(x) \cdot g(x)] = \lim_{x \to c} f(x) \cdot \lim_{x \to c} g(x)$

More Properties of Limit

4) $\lim_{x \to c} \dfrac{f(x)}{g(x)} = \dfrac{\lim_{x \to c} f(x)}{\lim_{x \to c} g(x)}$, provided $\lim_{x \to c} g(x) \neq 0$

5) $\lim_{x \to c}(af(x)) = a \lim_{x \to c} f(x)$

6) If r and s are integers, $s \neq 0$, then:

$$\lim_{x \to c}(f(x))^{r/s} = \left(\lim_{x \to c} f(x)\right)^{r/s}$$

provided $\left(\lim_{x \to c} f(x)\right)^{r/s}$ is real number.

Example 1

Find $\lim_{x \to 1}(2x+3)^2$.

$\lim_{x \to 1}(2x+3)^2 = \left(\lim_{x \to 1}(2x+3)\right)^2 =$

$\left(\lim_{x \to 1}(2x) + \lim_{x \to 1} 3\right)^2 = \left(2 \cdot \lim_{x \to 1} x + 3\right)^2 =$

$(2 \cdot 1 + 3)^2 = \boxed{25}$

Example 2

Find $\lim_{x \to 8} \dfrac{x^{2/3} - 2\sqrt{x}}{3 + \dfrac{24}{x}} = \dfrac{\lim_{x \to 8}\left(x^{2/3} - 2\sqrt{x}\right)}{\lim_{x \to 8}\left(3 + \dfrac{24}{x}\right)} =$

$\dfrac{\lim_{x \to 8} x^{2/3} - \lim_{x \to 8} 2\sqrt{x}}{\lim_{x \to 8} 3 + \lim_{x \to 8} \dfrac{24}{x}} = \dfrac{\left(\lim_{x \to 8} x\right)^{2/3} - 2\left(\lim_{x \to 8} x\right)^{1/2}}{\lim_{x \to 8} 3 + \dfrac{\lim_{x \to 8} 24}{\lim_{x \to 8} x}} =$

$\dfrac{8^{\frac{2}{3}} - 2 \cdot 8^{\frac{1}{2}}}{3 + \dfrac{24}{8}} = \dfrac{4 - 4\sqrt{2}}{6} = \boxed{\dfrac{2 - 2\sqrt{2}}{3}}$

For Polynomial and Rational Functions:

1. If $f(x) = a_n x^n + a_{n-1} x^{n-1} + \ldots + a_0$ is a polynomial function and c is a real number, then $\lim_{x \to c} f(x) = f(c) = a_n c^n + a_{n-1} c^{n-1} + \ldots + a_0$

Just substitute $x = c$.

2. If $f(x)$ and $g(x)$ are polynomials (or $\dfrac{f(x)}{g(x)}$ is a rational function) and c is a real number, then $\lim_{x \to c} \dfrac{f(x)}{g(x)} = \dfrac{f(c)}{g(c)}$.

Just substitute $x = c$.

7 Methods of Evaluating Limits

1. Direct Substitution Technique

This should always be the first choice.

Example Find $\lim_{x \to 2} \dfrac{5x - 2}{4x^2 + x - 7} = \dfrac{5 \cdot 2 - 2}{4 \cdot 2^2 + 2 - 7} = \dfrac{8}{11}$

EASY!!!

Once you use the Direct Substitution Technique and get zero in both numerator and denominator $\left(\dfrac{0}{0}\right)$, you have an Indeterminate form.

NOT EASY!!!   SO, SIMPLIFY!   HOW?

2. Cancellation Technique

Example

$\lim_{x \to 2} \dfrac{x^2 - 4}{x - 2} =$ Plug in $x = 2$ and get $\dfrac{0}{0}$

$\lim_{x \to 2} \dfrac{(x - 2)(x + 2)}{x - 2} =$ Factor!

$\lim_{x \to 2} (x + 2) = 4$ Plug in $x = 2$

Example

$\lim_{x \to 3} \dfrac{x^3 - 27}{x - 3} =$ Plug in $x = 3$ and get $\dfrac{0}{0}$

$\lim_{x \to 3} \dfrac{(x - 3)(x^2 + 3x + 9)}{x - 3} =$ Factor!

$\lim_{x \to 3} (x^2 + 3x + 9) =$ Plug in $x = 3$

$3^2 + 3 \cdot 3 + 9 = \boxed{27}$

3. Rationalizing Numerator or Denominator Technique

Example

$\lim_{x \to 9} \dfrac{x - 9}{\sqrt{x} - 3} =$ Plug in $x = 9$ and get $\dfrac{0}{0}$

$\lim_{x \to 9} \dfrac{x - 9}{\sqrt{x} - 3} \cdot \dfrac{\sqrt{x} + 3}{\sqrt{x} + 3} =$ Multiply both numerator and denominator by $\sqrt{x} + 3$

$\lim_{x \to 9} \dfrac{(x - 9)(\sqrt{x} + 3)}{(x - 9)} =$

$\lim_{x \to 9} (\sqrt{x} + 3) = \boxed{6}$ Plug in $x = 9$

Example

$\lim_{x \to 0} \dfrac{1 - \sqrt{1 + x}}{x} =$ Plug in $x = 0$ and get $\dfrac{0}{0}$

$\lim_{x \to 0} \dfrac{1 - \sqrt{1 + x}}{x} \cdot \dfrac{1 + \sqrt{1 + x}}{1 + \sqrt{1 + x}} =$ Multiply both numerator and denominator by $1 + \sqrt{1 + x}$

$\lim_{x \to 0} \dfrac{1 - (1 + x)}{x(1 + \sqrt{1 + x})} =$ Simplify

$\lim_{x \to 0} \dfrac{-x}{x(1 + \sqrt{1 + x})} =$

$\lim_{x \to 0} \dfrac{-1}{1 + \sqrt{1 + x}} = \boxed{-\dfrac{1}{2}}$ Plug in $x = 0$

4. Simplifying Complex Fraction Technique
Example

$\lim_{x \to 0} \dfrac{\dfrac{1}{x+3} - \dfrac{1}{3}}{x} =$ Plug in $x=0$ and get $\dfrac{0}{0}$

$\lim_{x \to 0} \dfrac{\dfrac{3-(x+3)}{(x+3) \cdot 3}}{x} =$ Simplify numerator

$\lim_{x \to 0} \dfrac{3-x-3}{x \cdot 3 \cdot (x+3)} =$

$\lim_{x \to 0} \dfrac{-1}{3 \cdot (x+3)} = \boxed{-\dfrac{1}{9}}$ Plug in $x=0$

5. Simplifying Trig. Fraction Technique
Example

$\lim_{x \to \frac{\pi}{2}} \dfrac{\cos x}{\cot x} =$ Plug in $x = \dfrac{\pi}{2}$ and get $\dfrac{0}{0}$

$\lim_{x \to \frac{\pi}{2}} \dfrac{\cos x}{\dfrac{\cos x}{\sin x}} =$ Simplify

$\lim_{x \to \frac{\pi}{2}} \sin x = \sin \dfrac{\pi}{2} = \boxed{1}$ Plug in $x = \dfrac{\pi}{2}$

6. Graphical Technique
Example

$\lim_{x \to 0} \dfrac{1-\cos 3x}{x} = ?$ Plug in $x=0$ and get $\dfrac{0}{0}$

Graph $y = \dfrac{1-\cos 3x}{x}$ and check what it is approaching to when $x \to 0$

$\lim_{x \to 0} \dfrac{1-\cos 3x}{x} = \boxed{0}$

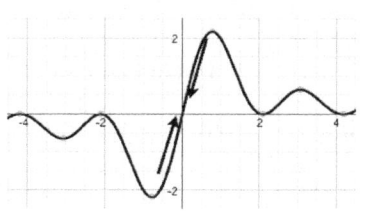

7. Evaluating One-Sided Limits Technique

Example. Find $\lim_{x \to 2} \dfrac{|x-2|}{x-2}$

$\lim_{x \to 2^+} \dfrac{|x-2|}{x-2} \Rightarrow \dfrac{|2.0001-2|}{2.0001-2} = \dfrac{|0.0001|}{0.0001} = \dfrac{0.0001}{0.0001} = \boxed{1}$

$\lim_{x \to 2^-} \dfrac{|x-2|}{x-2} \Rightarrow \dfrac{|1.9999-2|}{1.9999-2} = \dfrac{|-0.0001|}{-0.0001} = \dfrac{0.0001}{-0.0001} = \boxed{-1}$

$\lim_{x \to 2^+} \dfrac{|x-2|}{x-2} \neq \lim_{x \to 2^-} \dfrac{|x-2|}{x-2} \Rightarrow \lim_{x \to 2} \dfrac{|x-2|}{x-2}$ $\boxed{\text{DNE}}$

Example. Find $\lim_{x \to 1} f(x)$ if $f(x) = \begin{cases} 4-x^2, & x \leq 1 \\ 3x, & x > 1 \end{cases}$

$\lim_{x \to 1^-} f(x) \Rightarrow 4 - 0.999^2 \approx 3; \lim_{x \to 1^+} f(x) \Rightarrow 3 \cdot 1.001 \approx 3 \Rightarrow \lim_{x \to 1} f(x) = \boxed{3}$

If we cannot find a limit directly, we sometimes use

Sandwich Theorem:

If $f(x) \leq g(x) \leq h(x)$ for all $x \neq c$ in some interval about c, and $\lim_{x \to c} f(x) = \lim_{x \to c} h(x) = L$, then $\lim_{x \to c} g(x) = L$

 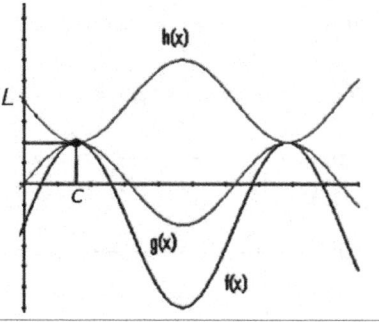

Using the Sandwich Theorem

Prove: $\lim_{x \to 0} x^3 \sin \dfrac{1}{x^3} = 0$

We know: $-1 \leq \sin \dfrac{1}{x^3} \leq 1$

$-1 \cdot x^3 \leq x^3 \cdot \sin \dfrac{1}{x^3} \leq 1 \cdot x^3$

$\lim_{x \to 0} (-1 \cdot x^3) \leq \lim_{x \to 0} \left(x^3 \cdot \sin \dfrac{1}{x^3} \right) \leq \lim_{x \to 0} (1 \cdot x^3)$

$0 \leq \lim_{x \to 0} \left(x^3 \sin \dfrac{1}{x^3} \right) \leq 0$

$\lim_{x \to 0} \left(x^3 \sin \dfrac{1}{x^3} \right) = 0$

1.3 Limits Involving Infinity. Horizontal and Vertical Asymptotes

Mathboat.com

Evaluating Limits of Rational Functions When x Approaches to Infinity

Example 1. Evaluate:

$$\lim_{x\to\infty}\frac{x^3-4x}{2x^4+5}=\lim_{x\to\infty}\frac{\dfrac{x^3}{x^4}-\dfrac{4x}{x^4}}{\dfrac{2x^4}{x^4}+\dfrac{5}{x^4}}=$$

Divide by the highest power of denominator

$$\lim_{x\to\infty}\frac{\dfrac{1}{x}-\dfrac{4}{x^3}}{2+\dfrac{5}{x^4}}=\frac{0-0}{2+0}=0$$

Evaluating limits when x approaches a constant, but $f(x)$ approaches to $+\infty$ or $-\infty$

Example 2a

Evaluate: $\lim\limits_{x\to 1}\dfrac{1}{(x-1)^2}$

Solution

As $x\to 1, (x-1)^2 > 0$ and approaches 0

$\lim\limits_{x\to 1^-}\dfrac{1}{(x-1)^2}=\infty$; $\lim\limits_{x\to 1^+}\dfrac{1}{(x-1)^2}=\infty$

$\lim\limits_{x\to 1}\dfrac{1}{(x-1)^2}=\infty$

Example 2b

Evaluate: $\lim\limits_{x\to 1}\dfrac{1}{x-1}$

Solution

As $x\to 1, (x-1)$ could be >0 or <0 and approaches 0

$\lim\limits_{x\to 1^+}\dfrac{1}{x-1}=\infty$ $\lim\limits_{x\to 1^-}\dfrac{1}{x-1}=-\infty$

Since left and right hand limits are not equal,

$\lim\limits_{x\to 1}\dfrac{1}{x-1}$ DNE

Example 2c

The graph is shown below. Find $\lim\limits_{x\to a} f(x)$ and $\lim\limits_{x\to b} f(x)$.

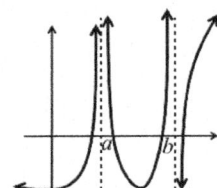

$\lim\limits_{x\to a^-} f(x)=\infty$

$\lim\limits_{x\to a^+} f(x)=\infty$

$\lim\limits_{x\to a} f(x)=\infty$

$\lim\limits_{x\to b^-} f(x)=\infty$

$\lim\limits_{x\to b^+} f(x)=-\infty$

$\lim\limits_{x\to b} f(x)$ DNE

Example 2d

Find $\lim\limits_{x\to 1}\dfrac{x^2-6}{(x-1)^3}$

$\lim\limits_{x\to 1}\dfrac{x^2-6}{(x-1)^3}\to\dfrac{-5}{0}\to$ Could be ∞, $-\infty$ or DNE

Take one-sided limits

$\lim\limits_{x\to 1^-}\dfrac{x^2-6}{(x-1)^3}\approx\dfrac{-5}{(.999-1)^3}=+\infty$

$\lim\limits_{x\to 1^+}\dfrac{x^2-6}{(x-1)^3}\approx\dfrac{-5}{(1.001-1)^3}=-\infty$ So, $\lim\limits_{x\to 1}\dfrac{x^2-6}{(x-1)^3}$ DNE

Horizontal Asymptotes

Definition: The line $y = L$ is a horizontal asymptote of the graph of a function $f(x)$ if $\lim_{x\to +\infty} f(x) = L$ or $\lim_{x\to -\infty} f(x) = L$.

Example 1. Find the horizontal asymptotes of $f(x) = \dfrac{2x}{\sqrt{x^2+1}}$

$\lim\limits_{x\to +\infty} \dfrac{2x}{\sqrt{x^2+1}} \Rightarrow \dfrac{2\cdot(10000)}{\sqrt{(10000)^2+1}} \Rightarrow \dfrac{20000}{10000} = 2$

$\lim\limits_{x\to -\infty} \dfrac{2x}{\sqrt{x^2+1}} \Rightarrow \dfrac{2\cdot(-10000)}{\sqrt{(-10000)^2+1}} \Rightarrow \dfrac{-20000}{10000} = -2$

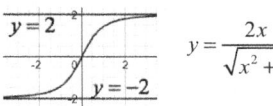

$y = \dfrac{2x}{\sqrt{x^2+1}}$

Horizontal Asymptotes:
$y = 2$
$y = -2$

Shortcut for finding Horizontal Asymptotes

Case 1: Power of Numerator = Power of Denominator

Find Horisontal Asymptote of $y = \dfrac{6x^2+3}{3x^2-7}$

$\lim\limits_{x\to\pm\infty} \dfrac{6x^2+3}{3x^2-7} = \lim\limits_{x\to\infty} \dfrac{\dfrac{6x^2}{x^2}+\dfrac{3}{x^2}}{\dfrac{3x^2}{x^2}-\dfrac{7}{x^2}} = \dfrac{6}{3} = 2$

Horizontal Asymptote: $y = 2$

Shortcut: Just divide coefficients of highest power into each other!

Do not use the shortcut if the function is not rational (as in Example 1)

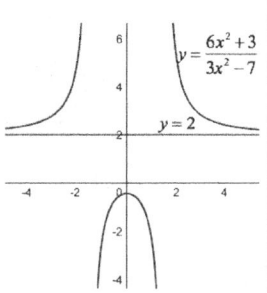

Shortcut for finding Horizontal Asymptotes

Case 2: Power of Numerator < Power of Denominator

Find Horisontal Asymptote of $y = \dfrac{x}{4x^2-7}$

$\lim\limits_{x\to\infty} \dfrac{x}{4x^2-7} = \lim\limits_{x\to\infty} \dfrac{\dfrac{x}{x^2}}{\dfrac{4x^2}{x^2}-\dfrac{7}{x^2}} = \dfrac{0}{4} = 0$

Horizontal Asymptote: $y = 0$

Shortcut: Horizontal asymptote will always be the x-axis

Shortcut for finding Horizontal Asymptotes

Case 3: Power of Numerator > Power of Denominator

Example: Find Horizontal Asymptote of $y = \dfrac{x^2+7}{x-3}$

First, we do long division:

```
         x + 3  R16
x - 3 ) x² + 7
        x² − 3x
        ───────
            3x + 7
            3x − 9
            ──────
                16
```

Then we find the limit ...

Case 3 Example Continued:

$\lim\limits_{x\to\pm\infty} \dfrac{x^2+7}{x-3} = \lim\limits_{x\to\infty}\left(x+3+\dfrac{16}{x-3}\right)$

When $x \to \infty$, $\dfrac{16}{x-3} \to 0$

The slant asymptote: $y = x+3$

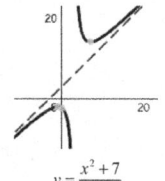

Shortcut: No horizontal asymptote. Find slant asymptote by long division

Vertical Asymptotes

The line $x = a$ is a vertical asymptote of the graph of a function $y = f(x)$ if $\lim\limits_{x\to a^+} f(x) = \pm\infty$ or $\lim\limits_{x\to a^-} f(x) = \pm\infty$

Example. Find vertical asymptote of $y = \dfrac{3}{x+2}$

The function approaches ∞ or $-\infty$ when $x \to -2$ from the right and from the left:

$\lim\limits_{x\to -2^+} \dfrac{3}{x+2} = \infty$, $\lim\limits_{x\to -2^-} \dfrac{3}{x+2} = -\infty$

So, vertical asymptote is $x = -2$

Shortcut: Set denominator = 0
$x + 2 = 0$
$x = -2$

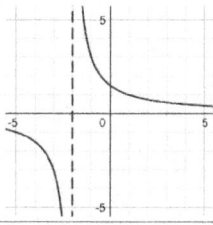

1.4 Continuous Functions

Mathboat.com

Continuity at the point

$f(x)$ is continuous at $x = c$ if:

1. $\lim\limits_{x \to c} f(x)$ exists

2. $f(c)$ exists

3. $\lim\limits_{x \to c} f(x) = f(c)$

Is function $f(x)$ continuous?

$\lim\limits_{x \to c} f(x)$ does not exist

$f(x)$ is not continuous at c

$f(c)$ does not exist

$f(x)$ is not continuous at c

Is function $f(x)$ continuous?

$\lim\limits_{x \to c} f(x) \neq f(c)$

$f(x)$ is not continuous at c

$\lim\limits_{x \to c} f(x) = f(c)$

$f(x)$ is continuous at c

Example 1

If $x > -2$, determine if the function $f(x) = \begin{cases} \dfrac{x^4 - 16}{x^2 - 4} & \text{if } x \neq 2 \\ 5 & \text{if } x = 2 \end{cases}$

is continuous at $x = 2$. If $f(x)$ is not continuous at $x = 2$, state the reason.

$f(x)$ is continuous if $\lim\limits_{x \to c} f(x)$ exists, $f(c)$ exists, and $\lim\limits_{x \to c} f(x) = f(c)$.

$\lim\limits_{x \to 2} f(x) = \lim\limits_{x \to 2} \dfrac{(x^2 - 4)(x^2 + 4)}{(x^2 - 4)} = \lim\limits_{x \to 2}(x^2 + 4) = 4 + 4 = 8$, but $f(2) = 5$

$\lim\limits_{x \to 2} f(x) \neq f(2)$.

$f(x)$ is not continuous.

Example 2.

Determine if the function $f(x) = \begin{cases} x^2 & \text{if } x < 1 \\ 1 & \text{if } x = 1 \\ 2x - 1 & \text{if } x > 1 \end{cases}$

is continuous at $x = 1$. If $f(x)$ is not continuous at $x = 1$, state the reason.

$f(x)$ is continuous if $\lim\limits_{x \to c} f(x)$ exists, $f(c)$ exists, and $\lim\limits_{x \to c} f(x) = f(c)$.

$\left. \begin{array}{l} \lim\limits_{x \to 1^-} f(x) = 1^2 = 1 \\ \lim\limits_{x \to 1^+} f(x) = 2 \cdot 1 - 1 = 1 \end{array} \right\} \Rightarrow \lim\limits_{x \to 1} f(x) = 1$

$f(1) = 1 \Rightarrow \lim\limits_{x \to 1} f(x) = f(1)$

Therefore, $f(x)$ is continuous at $x = 1$.

Infinite Discontinuity

$f(x) \to \infty$ or $-\infty$ as $x \to c$.
$\lim_{x \to c} f(x)$ DNE and $f(c)$ DNE

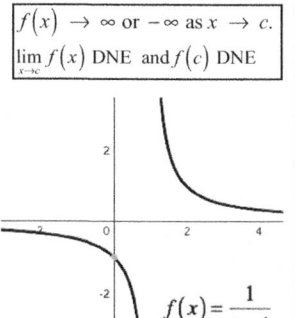

$f(x) = \dfrac{1}{x-1}$

$f(x) \to \infty$ or $-\infty$ as $x \to 1$.
$\lim_{x \to 1} f(x)$ DNE and $f(1)$ DNE

Jump Discontinuity

$\lim_{x \to c^-} f(x) \neq \lim_{x \to c^+} f(x)$ so $\lim_{x \to c} f(x)$ DNE
$f(x)$ approaches different values from the left of c and the right of c

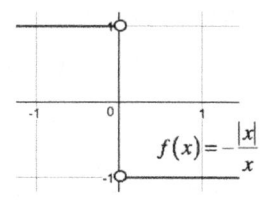

$f(x) = -\dfrac{|x|}{x}$

$\lim_{x \to 0^-} f(x) = 1, \lim_{x \to 0^+} f(x) = -1$
$\lim_{x \to 0^-} f(x) \neq \lim_{x \to 0^+} f(x)$ so $\lim_{x \to 0} f(x)$ DNE

Removable Discontinuity

| $f(c)$ is undefined | or $\lim_{x \to c} f(x) \neq f(c)$ |

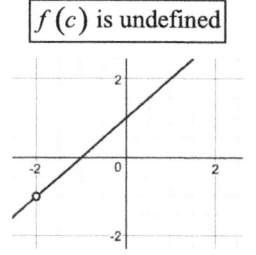 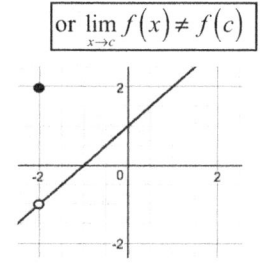

$f(x) = \dfrac{x^2+3x+2}{x+2} = \dfrac{\cancel{(x+2)}(x+1)}{\cancel{x+2}}$
$= x+1$

$f(-2)$ is undefined

$f(x) = \begin{cases} \dfrac{x^2+3x+2}{x+2}, & \text{if } x \neq -2 \\ 2, & \text{if } x = -2 \end{cases}$

$\lim_{x \to -2} f(x) \neq f(-2)$

Continuity on a Closed Interval

The function f is **continuous on** $[a,b]$ if it is continuous on (a,b) and if
$\lim_{x \to a^+} f(x) = f(a)$ and $\lim_{x \to b^-} f(x) = f(b)$

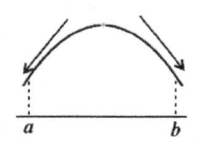

Example 3. Sketch the graph of $f(x) = \sqrt{16-x^2}$, and prove that f is continuous on the closed interval $[-4,4]$.
At $x = c, -4 < c < 4$:
$\lim_{x \to c} f(x) = \lim_{x \to c} \sqrt{16-x^2} = \sqrt{16-c^2} = f(c) \Rightarrow f(x)$ is continuous on $(-4,4)$
Check endpoints
$\lim_{x \to -4^+} f(x) = \lim_{x \to -4^+} \sqrt{16-x^2} = \sqrt{16-(-3.99)^2} = 0 = f(-4)$
$\lim_{x \to 4^-} f(x) = \lim_{x \to 4^-} \sqrt{16-x^2} = \sqrt{16-(3.99)^2} = 0 = f(4)$
Thus, f is continuous on $[-4,4]$

$y = \sqrt{16-x^2}$

Intermediate Value Theorem

If f is continuous on a closed interval $[a,b]$ and if w is any number between $f(a)$ and $f(b)$, then there is at least one number c in $[a,b]$ such that $f(c)=w$.

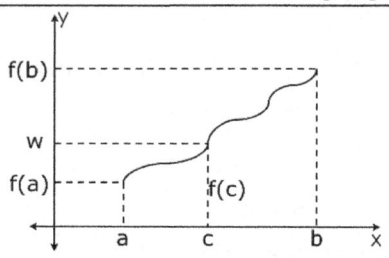 

OR: For any function f that is continuous over the interval $[a,b]$, the function will take any value between $f(a)$ and $f(b)$ over the interval

Example 4

Does $f(x) = x^5 + 3x^4 - 5x^3 + 2x - 4$ have a zero between $x = 1$ and $x = 2$?

Use Intermediate Value Theorem:

If f is continuous on a closed interval [a,b] and if w is any number between f(a) and f(b), then there is at least one number c in [a,b] such that f(c)=w.

$f(x)$ is a polynomial, Continuous

$f(1) = 1+3-5+2-4 = -3 < 0 \quad (a=1)$
$-3 < 0 < 40 \quad (w=0)$
$f(2) = 32+48-40+4-4 = 40 > 0 \quad (b=2)$

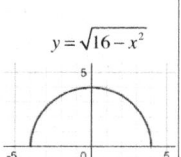

There is at least one real number **c** between **1** and **2** such that f(**c**)=**0**.

Example 5

Verify the Intermediate Value Theorem for $f(x) = x^2 - 2x$ on the interval $[1,4]$

$f(x)$ is a polynomial, continuous
$f(1) = 1-2 = -1, \quad a=1$
$f(4) = 16-8 = 8, \quad b=4$
If $-1 \leq w \leq 8$, then there is at least one c in $[1,4]$ such that $w = c^2 - 2c \quad \Leftrightarrow c^2 - 2c - w = 0$

$c = \dfrac{2 \pm \sqrt{4-4 \cdot 1 \cdot (-w)}}{2} = \dfrac{2 \pm \sqrt{4+4w}}{2} = 1 \pm \sqrt{1+w}.$

Is $c = 1 - \sqrt{1+w}$ possible? NO, for $-1 \leq w \leq 8$: $c = 1-\sqrt{1+w} \notin [1,4]$ (try $w=0$)
Is $c = 1 + \sqrt{1+w}$ possible? YES, for $-1 \leq w \leq 8$: $c = 1+\sqrt{1+w} \in [1,4]$
$\Rightarrow$ If $-1 \leq w \leq 8$, there is at least one number c in $[1,4]$ such that $f(c) = w$

If f is continuous on a closed interval [a,b] and if **w** is any number between f(a) and f(b), then there is **at least** one number **c** in [a,b] such that f(**c**)=**w**.

1.5 Sum, Product, Quotient and Composition of Limits

Mathboat.com

Properties of limits

Recall: If $\lim\limits_{x\to c} f(x)$, $\lim\limits_{x\to c} g(x)$, c are real numbers, then:

$$\lim_{x\to c}[f(x)+g(x)] = \lim_{x\to c} f(x) + \lim_{x\to c} g(x)$$

$$\lim_{x\to c}[f(x)-g(x)] = \lim_{x\to c} f(x) - \lim_{x\to c} g(x)$$

$$\lim_{x\to c}[f(x)\cdot g(x)] = \lim_{x\to c} f(x) \cdot \lim_{x\to c} g(x)$$

$$\lim_{x\to c}\frac{f(x)}{g(x)} = \frac{\lim_{x\to c} f(x)}{\lim_{x\to c} g(x)}, \text{ provided } \lim_{x\to c} g(x) \neq 0$$

These are only true IF EACH LIMIT EXISTS and THEY ARE REAL NUMBERS and do not create an indeterminate form.

What if some limits don't exist?

For example, if $\lim\limits_{x\to c} f(x)$ does not exist, then
$$\lim_{x\to c}[f(x)\cdot g(x)] \neq \lim_{x\to c} f(x) \cdot \lim_{x\to c} g(x).$$

So try to find right and left hand limits of **entire product**:
$$\lim_{x\to c^+}[f(x)\cdot g(x)] \text{ and } \lim_{x\to c^-}[f(x)\cdot g(x)].$$

If they are equal to L, your answer is L.
If they are not equal, your answer is DNE.
Later on, we will learn methods to deal with limits if they create indeterminate forms 0/0 etc.

Example 1 Compute $\lim\limits_{x\to -1}(f(x)g(x))$

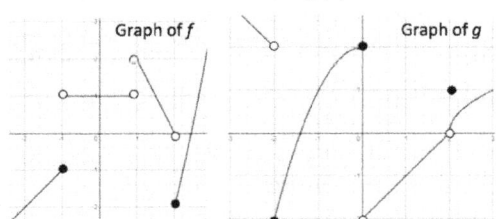

$\lim\limits_{x\to -1} f(x)$ DNE, so $\lim\limits_{x\to -1}(f(x)g(x))$ cannot be written as product of limits
⇒ find Left and Right hand limits of the entire product.

Left hand limit: $\lim\limits_{x\to -1^-}(f(x)g(x)) = \lim\limits_{x\to -1^-} f(x) \cdot \lim\limits_{x\to -1^-} g(x) = -1 \cdot 1 = \boxed{-1}$

Right hand limit: $\lim\limits_{x\to -1^+}(f(x)g(x)) = \lim\limits_{x\to -1^+} f(x) \cdot \lim\limits_{x\to -1^+} g(x) = 1 \cdot 1 = \boxed{1}$

$\lim\limits_{x\to -1^-}(f(x)g(x)) \neq \lim\limits_{x\to -1^+}(f(x)g(x)) \Rightarrow \boxed{\lim\limits_{x\to -1}(f(x)g(x)) \text{ DNE}}$

Example 2 Compute $\lim\limits_{x\to 2}(f(x)g(x))$

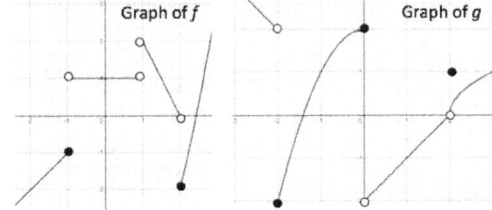

$\lim\limits_{x\to 2} f(x)$ DNE, so $\lim\limits_{x\to 2}(f(x)g(x))$ cannot be written as product of limits
⇒ find Left and Right hand limits of the entire product.

Left hand limit: $\lim\limits_{x\to 2^-}(f(x)g(x)) = \lim\limits_{x\to 2^-} f(x) \cdot \lim\limits_{x\to 2^-} g(x) = 0 \cdot 0 = \boxed{0}$

Right hand limit: $\lim\limits_{x\to 2^+}(f(x)g(x)) = \lim\limits_{x\to 2^+} f(x) \cdot \lim\limits_{x\to 2^+} g(x) = -2 \cdot 0 = \boxed{0}$

$\lim\limits_{x\to 2^-}(f(x)g(x)) = \lim\limits_{x\to 2^+}(f(x)g(x)) = 0 \Rightarrow \boxed{\lim\limits_{x\to 2}(f(x)g(x)) = 0}$

Example 3

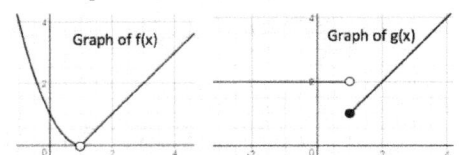

Compute $\lim\limits_{x\to 1}\dfrac{f(x)}{g(x)}$.

$\lim\limits_{x\to 1} g(x)$ DNE so we cannot write $\lim\limits_{x\to 1}\dfrac{f(x)}{g(x)}$ as a quotient of limits.

Left hand limit of the entire quotient: $\lim\limits_{x\to 1^-}\dfrac{f(x)}{g(x)} = \dfrac{\lim\limits_{x\to 1^-} f(x)}{\lim\limits_{x\to 1^-} g(x)} = \dfrac{0}{2} = \boxed{0}$

Right hand limit of the entire quotient: $\lim\limits_{x\to 1^+}\dfrac{f(x)}{g(x)} = \dfrac{\lim\limits_{x\to 1^+} f(x)}{\lim\limits_{x\to 1^+} g(x)} = \dfrac{0}{1} = \boxed{0}$

Left and right hand limits are both 0 so $\lim\limits_{x\to 1}\dfrac{f(x)}{g(x)} = \boxed{0}$

Composition of Limits

How do we compute $\lim_{x \to a} f(g(x))$?

It seems that if $\lim_{x \to a} g(x) = L$, $\lim_{y \to L} f(y) = M$,
then $\lim_{x \to a} f(g(x)) = M$.

However, this is not strictly true:

If $\lim_{x \to a} g(x) = L$, $\lim_{y \to L} f(y) = M$ **and**
$f(x)$ is CONTINUOUS at L,
then $\lim_{x \to a} f(g(x)) = M$.

Example 4

Find the value of these limits:

a) $\lim_{x \to -5} f(g(x)) = ?$

As $x \to -5$, $g(x) \to 2$ from both sides

So $\lim_{x \to -5} f(g(x)) = \lim_{y \to 2} f(y) = \boxed{2}$

b) $\lim_{x \to 3} g(f(x)) = ?$

As $x \to 3$, $f(x) \to 4$ from both sides

So $\lim_{x \to 3} g(f(x)) = \lim_{y \to 4} g(y) = \boxed{-3}$

c) $\lim_{x \to 4} g(f(x)) = ?$

As $x \to 4$, $f(x) \to 2$ from both sides

So $\lim_{x \to 4} g(f(x)) = \lim_{y \to 2} g(y)$ $\boxed{DNE}$

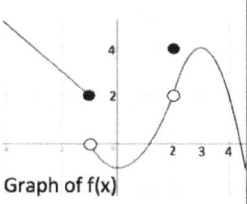
Graph of f(x)

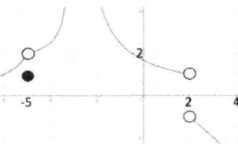
Graph of g(x)

Example 5 $\lim_{x \to 1} f(g(x)) = ?$

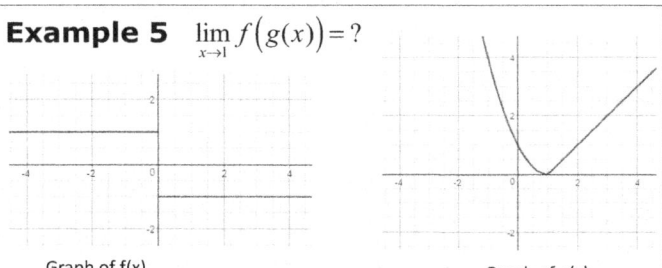

Graph of f(x) Graph of g(x)

$\lim_{x \to 1} g(x) = 0$, but $f(x)$ is not continuous at 0.

As $x \to 1$, $g(x) \to 0$, but $g(x)$ is always greater than 0.

Thus as $x \to 1$, $g(x) \to 0^+$ (from above).

So $\lim_{x \to 1} f(g(x)) = \lim_{y \to 0^+} f(y) = \boxed{-1}$

Example 6

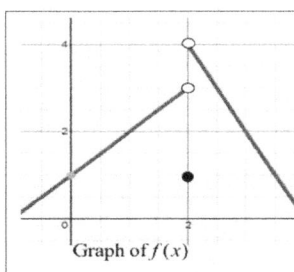
Graph of $f(x)$

The function $f(x)$ is given by the graph on the left.

The function $g(x)$ is given by the formula: $g(x) = (x-1)^2 + 2$.

$\lim_{x \to 1} f(g(x)) = ?$

As $x \to 1$, $(x-1)^2 > 0$, so $(x-1)^2 \to 0^+$

$\Rightarrow \lim_{x \to 1} g(x) = \lim_{x \to 1} ((x-1)^2 + 2) = 2^+$ or "2 from above"

So $\lim_{x \to 1} f(g(x)) = \lim_{y \to 2^+} f(y) = \boxed{4}$

Example 7

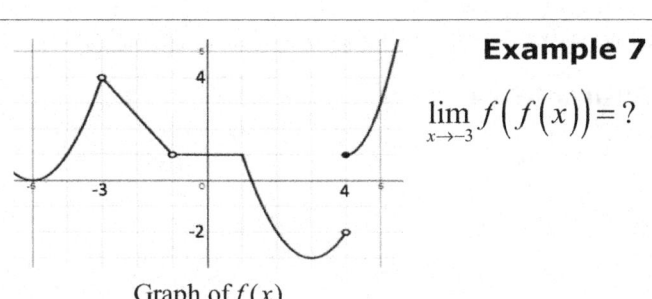
Graph of $f(x)$

$\lim_{x \to -3} f(f(x)) = ?$

$\lim_{x \to -3} f(x) = 4^-$ (from below)

$\lim_{x \to -3} f(\underbrace{f(x)}_{y}) = \lim_{y \to 4^-} f(y) = \boxed{-2}$

Example 8

Compute $\lim_{x \to \frac{\pi}{2}} f(\sin x)$.

When $x \to \frac{\pi}{2}$, $\sin x \to 1$, but $\sin x < 1$,

so $\sin x$ approaches 1^- (from below),

$\Rightarrow \lim_{x \to \frac{\pi}{2}} \sin x = 1^-$

so $\lim_{x \to \frac{\pi}{2}} f(\sin x) = \lim_{y \to 1^-} f(y) = \boxed{0}$

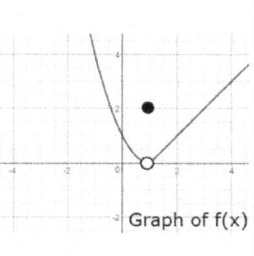
Graph of f(x)

Example 9

Compute $\lim_{x\to\frac{\pi}{2}} f(\sin x)$.

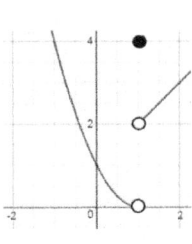
Graph of f(x)

What if there is a jump discontinuity at $x = 1$?

As $x \to \frac{\pi}{2}$, $\sin x \to 1$, but $\sin x$ is always less than 1.

Thus as $x \to \frac{\pi}{2}$, $\sin x \to 1^-$.

So $\lim_{x\to\frac{\pi}{2}} f(\sin x) = \lim_{y\to 1^-} f(y) = \boxed{0}$.

Example 10

Compute $\lim_{x\to 0^-} f(x+2)$ and $\lim_{x\to 0^+} f(x+2)$

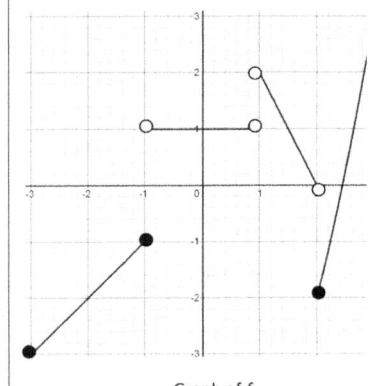
Graph of f

As $x \to 0^-$, $x+2 \to 2^-$
$\Rightarrow \lim_{x\to 0^-} f(x+2) = \lim_{y\to 2^-} f(y) = \boxed{0}$

As $x \to 0^+$, $x+2 \to 2^+$
$\Rightarrow \lim_{x\to 0^+} f(x+2) = \lim_{y\to 2^+} f(y) = \boxed{-2}$

Example 11

Compute $\lim_{x\to -1^-} f(x^2)$ and $\lim_{x\to -1^+} f(x^2)$

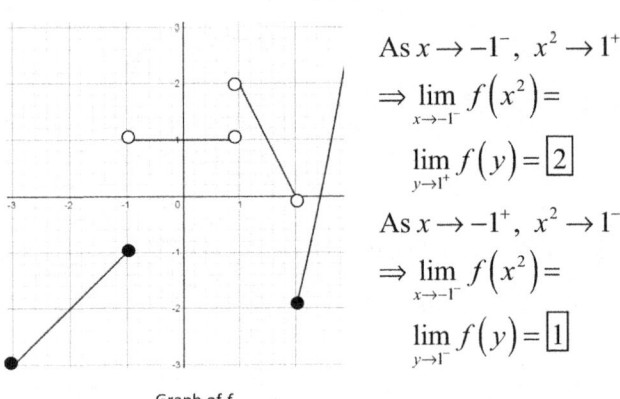
Graph of f

As $x \to -1^-$, $x^2 \to 1^+$
$\Rightarrow \lim_{x\to -1^-} f(x^2) = \lim_{y\to 1^+} f(y) = \boxed{2}$

As $x \to -1^+$, $x^2 \to 1^-$
$\Rightarrow \lim_{x\to -1^+} f(x^2) = \lim_{y\to 1^-} f(y) = \boxed{1}$

Example 12

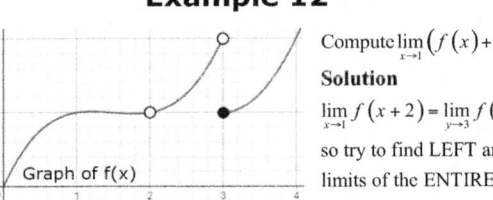
Graph of f(x)

Compute $\lim_{x\to 1}(f(x)+f(x+2))$.

Solution

$\lim_{x\to 1} f(x+2) = \lim_{y\to 3} f(y) \Rightarrow DNE$, so try to find LEFT and RIGHT hand limits of the ENTIRE SUM

Since $\lim_{x\to 1^-} f(x) = 1$, $\lim_{x\to 1^-} f(x+2) = \lim_{y\to 3^-} f(y) = 2$, then
LEFT HAND LIMIT $= \lim_{x\to 1^-}(f(x)+f(x+2)) = 1+2 = \boxed{3}$

Since $\lim_{x\to 1^+} f(x) = 1$, $\lim_{x\to 1^+} f(x+2) = \lim_{y\to 3^+} f(y) = 1$, then
RIGHT HAND LIMIT $= \lim_{x\to 1^+}(f(x)+f(x+2)) = 1+1 = \boxed{2}$

LEFT and RIGHT hand limits are DIFFERENT, so $\lim_{x\to 1}(f(x)+f(x+2))$ $\boxed{DNE}$.

Example 13

Compute $\lim_{x\to 2}(f(x-1)g(x+2))$.

$\lim_{x\to 2}(g(x+2))$ DNE
$\Rightarrow \lim_{x\to 2}(f(x-1)g(x+2))$ cannot be written as product of limits.

Right hand limit: $\lim_{x\to 2^+}(f(\underbrace{x-1}_{y})\underbrace{g(x+2)}_{z}) =$

$\lim_{y\to 1^+} f(y) \cdot \lim_{z\to 4^+} g(z) = 0 \cdot 1 = \boxed{0}$

Left hand limit: $\lim_{x\to 2^-}(f(\underbrace{x-1}_{y})\underbrace{g(x+2)}_{z}) =$

$\lim_{y\to 1^-} f(y) \cdot \lim_{z\to 4^-} g(z) = 0 \cdot 3 = \boxed{0}$

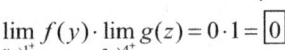
Graph of f

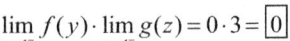
Graph of g

Left and right hand limits are both $0 \Rightarrow \boxed{\lim_{x\to 2}(f(x-1)g(x+2)) = 0}$

Finding $\lim_{x\to c} f(g(x))$ if $g(x) = a$ (horizontal) around $x = c$ and f discontinues at a.

Example 14

Compute $\lim_{x\to 1} f(g(x))$.

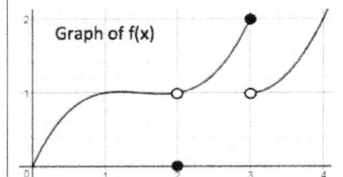
Graph of f(x)

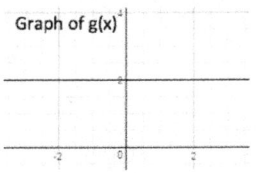
Graph of g(x)

$\lim_{x\to 1} g(x) = 2$, $\lim_{y\to 2} f(y) = 1$, so $\boxed{\text{IS } \lim_{x\to 1} f(g(x)) = 1 \text{ ???}}$ **NO !!!**, since

1) $f(x)$ is not continuous at $x = 2$ and
2) $g(x) = 2$ for all x or at least $\underline{g(x) = 2}$ on an interval around $x = 1$

So, $\lim_{x\to 1} f(g(x)) = \lim_{x\to 1} f(2) = \lim_{x\to 1} 0 = \boxed{0}$

2.1 Tangent Lines and Rates of Change. Average and Instantaneous Velocity.

Tangent Lines

- The tangent line to the graph of f is a line that touches the graph at one isolated point and could possibly intersect it again at another point.

- Slope of the graph of f at point c is the slope of the tangent line at point c

Yes!

Yes!

not a tangent, 2 pts of intersection.

Let's first find the Slope of a secant line:

$$\text{Slope of secant line} = \frac{f(x) - f(c)}{x - c}$$

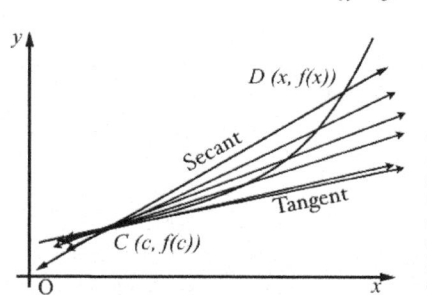

Let's pick a point closer and closer to point C and draw a secant.
Notice that secants are getting closer and closer to tangent line at point C and almost coincide with it.

Slope of tangent line at pt C ≈ Slope of Secant Line when x is approaching to C

Slope of tangent line at pt C ≈ Slope of Secant Line when x is approaching to C

Slope of tangent line $m_c = \lim\limits_{x \to c} \dfrac{f(x) - f(c)}{x - c}$

Let $x - c = h$ So, $x = c + h$

$x \to c \Rightarrow x - c \to 0 \Rightarrow h \to 0$

$$m_c = \lim_{h \to 0} \frac{f(c+h) - f(c)}{h}$$

Replace every c with x

$$m_x = \lim_{h \to 0} \frac{f(x+h) - f(x)}{h}$$

To Find the Equation of the Tangent Line at $x=c$:

1. Find the slope of tangent at any point x:

$$m_x = \lim_{h \to 0} \frac{f(x+h) - f(x)}{h}$$

2. Plug in $x = c$ into m_x to find the slope m_c at the point $(c, f(c))$.

Or (instead of 1 and 2) just use the formula for the slope at the point $x = c$: $m_c = \lim\limits_{x \to c} \dfrac{f(x) - f(c)}{x - c}$

3. Substitute coordinates $(c, f(c))$ and slope m_c into the point-slope equation of a line:

$$y - y_0 = m(x - x_0)$$

$$y - f(c) = m_c(x - c)$$

Example 1

Find the equation of the tangent line to the curve $f(x) = x^2 + 3x + 7$ at the point $(-2, 5)$.

1. Find the slope of tangent at any point x:

$$m_x = \lim_{h \to 0} \frac{f(x+h) - f(x)}{h} = \lim_{h \to 0} \frac{(x+h)^2 + 3(x+h) + 7 - (x^2 + 3x + 7)}{h}$$

$$= \lim_{h \to 0} \frac{\cancel{x^2} + 2xh + h^2 + \cancel{3x} + 3h + \cancel{7} - \cancel{x^2} - \cancel{3x} - \cancel{7}}{h} = \lim_{h \to 0} \frac{2xh + h^2 + 3h}{h}$$

$$= \lim_{h \to 0} \frac{\cancel{h}(2x + h + 3)}{\cancel{h}} = \lim_{h \to 0}(2x + h + 3) \Rightarrow \boxed{m_x = 2x + 3}$$

2. Plug in $x = -2$ into m_x to find the slope at the point $(-2, f(-2))$: $m = 2(-2) + 3 = \boxed{-1}$

Example 1 Continued

Or instead of 1. and 2.,

Use formula of slope at the point $x = c$:

$$m_c = \lim_{x \to -2} \frac{f(x)-f(c)}{x-c} = \lim_{x \to -2} \frac{f(x)-f(-2)}{x-(-2)}$$

$$= \lim_{x \to -2} \frac{(x^2+3x+7)-5}{x-(-2)} = \lim_{x \to -2} \frac{x^2+3x+2}{x+2}$$

$$= \lim_{x \to -2} \frac{(x+2)(x+1)}{x+2} = \lim_{x \to -2}(x+1) = \boxed{-1}$$

3. Plug in to equation of the line: $y - y_0 = m(x - x_0)$

$y - 5 = -1(x+2) \Rightarrow \boxed{y = -x + 3}$

Average Velocity and Rate of Change

The average velocity of a body moving along a line between times a and $a+h$ is:

$$V_{av} = \frac{\text{change in distance}}{\text{change in time}} \quad \text{or} \quad V_{av} = \frac{S(a+h)-S(a)}{h}$$

Average rate of change of $y = f(x)$ on $[a, a+h]$ is: $\dfrac{f(a+h)-f(a)}{h}$

Example. The position function of a point P moving along the line is $s(t) = 5t^3 - 2t^2 + 8$. Find the average velocity of P on the interval $[1, 1.4]$

$$V_{av} = \frac{s(a+h)-s(a)}{h} = \frac{s(1.4)-s(1)}{1.4-1} = \frac{(5 \cdot 1.4^3 - 2 \cdot 1.4^2 + 8)-(5 \cdot 1^3 - 2 \cdot 1^2 + 8)}{.4} = \boxed{17}$$

Instantaneous Velocity and Rate of Change

The instantaneous velocity of a body moving along a line between times a and $a+h$ is:

$$V = \lim_{h \to 0}(\text{Average Velocity}) \quad \boxed{V = \lim_{h \to 0} \frac{S(a+h)-S(a)}{h}}$$

Instantaneous Rate of Change of $y = f(x)$ on $[a, a+h]$ is:

$$\lim_{h \to 0}(\text{Average Rate of Change}) = \lim_{h \to 0} \frac{f(a+h)-f(a)}{h}$$

Example 2

A sandbag is dropped from a balloon 480 feet above the ground. The distance s(t) in feet from the ground to the sandbag after t seconds is given by:

$$s(t) = -16t^2 + 480$$

Find the velocity of the sandbag in ft/sec at:

(a) t = a sec   (b) t = 3 sec   (c) the instant it strikes the ground

(a) $V_a = \lim_{h \to 0} \dfrac{s(a+h)-s(a)}{h} = \lim_{h \to 0} \dfrac{[-16(a+h)^2 + 480]-(-16a^2 + 480)}{h}$

$= \lim_{h \to 0} \dfrac{-16(a^2 + 2ah + h^2) + 480 + 16a^2 - 480}{h} = \lim_{h \to 0} \dfrac{-32ah - 16h^2}{h}$

$= \lim_{h \to 0} \dfrac{h(-32a - 16h)}{h} = \lim_{h \to 0}(-32a - 16h) = -32a \, ft/sec$

Example 2 Continued

$\boxed{s(t) = -16t^2 + 480}$ $\boxed{V_a = -32a \, ft/sec}$

(b) Find the velocity in ft/sec at $t = 3$ sec

$V_t = -32t \, ft/sec$

$V_3 = -32 \cdot 3 \, ft/sec = -96 \, ft/sec$

Example 2 Continued

$\boxed{s(t) = -16t^2 + 480}$ $\boxed{V_a = -32a \, ft/sec}$

(c) Find the velocity in ft/sec at the instant it hits the ground

$s(t) = -16t^2 + 480 = 0$

$-16t^2 = -480$

$t^2 = \dfrac{-480}{-16} = 30$ $t = \pm\sqrt{30} \Rightarrow t = \sqrt{30}$

$V_{\sqrt{30}} = -32 \cdot \sqrt{30} \, ft/sec$

2.2 Definition of Derivative.

Mathboat.com

Definition of Derivative

The derivative of function f is function f' defined by
$$f'(x) = \lim_{h \to 0} \frac{f(x+h) - f(x)}{h}$$ provided the limit exists

Alternative definition of derivative:
$$f'(a) = \lim_{x \to a} \frac{f(x) - f(a)}{x - a}$$

If $f'(x)$ exists *we say f* is differentiable at x or f has derivative at x.

$f'(x)$ exists means $\lim_{h \to 0} \frac{f(x+h) - f(x)}{h}$ exists.

$\lim_{h \to 0} \frac{f(x+h) - f(x)}{h}$ exists if $\underbrace{\lim_{h \to 0^+} \frac{f(x+h) - f(x)}{h}}_{\text{right-hand derivative at } x} = \underbrace{\lim_{h \to 0^-} \frac{f(x+h) - f(x)}{h}}_{\text{left-hand derivative at } x}$

$\lim_{h \to 0} \frac{f(x+h) - f(x)}{h}$ exists if $\underbrace{\lim_{h \to 0^+} \frac{f(x+h) - f(x)}{h}}_{\text{right-hand derivative at } x} = \underbrace{\lim_{h \to 0^-} \frac{f(x+h) - f(x)}{h}}_{\text{left-hand derivative at } x}$

Example 1

If $f(x) = |x - 4|$, show that f is not differentiable at $x = 4$.

Prove that left-hand and right-hand derivatives are not equal:

Right-hand derivative at $x = 4$ is: $\lim_{h \to 0^+} \frac{f(x+h) - f(x)}{h} =$

$\lim_{h \to 0^+} \frac{f(4+h) - f(4)}{h} = \lim_{h \to 0^+} \frac{|4+h-4| - |4-4|}{h} = \lim_{h \to 0^+} \frac{|h|}{h} = \boxed{1}$

Left-hand derivative at $x = 4$ is: $\lim_{h \to 0^-} \frac{f(x+h) - f(x)}{h} =$

$\lim_{h \to 0^+} \frac{f(4+h) - f(4)}{h} = \lim_{h \to 0^-} \frac{|h|}{h} = \boxed{-1}$ $\boxed{1 \neq -1}$

$\boxed{f(x) \text{ is not differentiable at } x = 4}$

Example 2

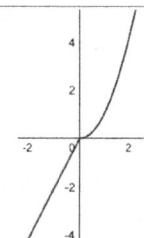

Given function $f(x) = \begin{cases} 2x & \text{if } x \leq 0 \\ x^2 & \text{if } x > 0 \end{cases}$

Is $f(x)$ differentiable at $x = 0$?

Check if $f'(x)$ exists at $x = 0$

Check if left- and right-hand derivatives are equal at $x=0$.

Left-hand derivative at $x = 0$: $\lim_{h \to 0} \frac{2(0+h) - (2 \cdot 0)}{h} = \lim_{h \to 0} \frac{2h}{h} = 2$

Right-hand derivative at $x = 0$: $\lim_{h \to 0^+} \frac{(0+h)^2 - 0^2}{h} = \lim_{h \to 0^+} h = 0$

At $x = 0$, the right hand and left hand derivatives are not equal, so $f(x)$ is not differentiable at $x = 0$.

Applications of Derivative

$f'(a) =$ slope of tangent line to $f(x)$ at $(a, f(a))$

$f'(a) =$ instantaneous rate of change of $f(x)$ with respect to x at $x = a$

Notations of Derivative

$$f'(x) = y' = \frac{dy}{dx} = \frac{d}{dx} f(x)$$

Why quotient ? Next slide...

Derivatives of the higher order

$y', y'', y''', \ldots, y^{(n)}, \ldots$ or $\frac{dy}{dx}, \frac{d^2 y}{dx^2}, \frac{d^3 y}{dx^3}, \ldots, \frac{d^n y}{dx^n}, \ldots$

Why $\frac{dy}{dx}$?

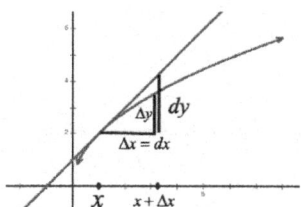

Slope of tangent $=$ quotient $\frac{dy}{dx}$

Derivative $= f'(x)$
$= \lim_{\Delta x \to 0} \frac{f(x + \Delta x) - f(x)}{\Delta x}$
$\approx \frac{\Delta y}{\Delta x}$ if $\Delta x \to 0$

When $\Delta x \to 0$: $\Delta y \to dy$ and $\Delta x = dx$, then quotient $\frac{dy}{dx} \approx \frac{\Delta y}{\Delta x} \approx f'(x)$

quotient of dy and $dx = \frac{dy}{dx} =$ symbol for y'

Example 3

Given: $f(x) = 3x^2 - 10x + 5$ a) Find $f'(x)$

$f'(x) = \lim_{h \to 0} \dfrac{f(x+h) - f(x)}{h}$

$= \lim_{h \to 0} \dfrac{\left(3(x+h)^2 - 10(x+h) + 5\right) - \left(3x^2 - 10x + 5\right)}{h}$

$= \lim_{h \to 0} \dfrac{\left(\cancel{3x^2} + 6xh + 3h^2 \cancel{-10x} - 10h \cancel{+5}\right) \cancel{-3x^2} + \cancel{10x} \cancel{-5}}{h}$

$= \lim_{h \to 0} \dfrac{6xh + 3h^2 - 10h}{h} = \lim_{h \to 0} \dfrac{h(6x + 3h - 10)}{h}$

$= \lim_{h \to 0}(6x + 3h - 10) = \boxed{6x - 10}$

Example 3 Continued

Given: $f(x) = 3x^2 - 10x + 5$ From part (a): $f'(x) = 6x - 10$

b) Find the domain of $f'(x)$

Since $f'(x) = 6x - 10$ is a line, the derivative exists for $\boxed{\text{all real } x}$

c) Find the slope of the tangent line to $f(x)$ at the point $(2, -3)$.

$f'(2) = 6 \cdot 2 - 10 = \boxed{2}$

d) Find the point on $f(x)$ at which tangent line is horizontal

Tangent line is horizontal if slope $= 0$

$f'(x) = 0 \Rightarrow 6x - 10 = 0 \Rightarrow x = \dfrac{5}{3}$

Plug into $f(x): f\left(\dfrac{5}{3}\right) = 3\left(\dfrac{5}{3}\right)^2 - 10\left(\dfrac{5}{3}\right) + 5 = -3\dfrac{1}{3} \Rightarrow \boxed{\left(\dfrac{5}{3}, -3\dfrac{1}{3}\right)}$

Differentiability on a Closed Interval

A function f is differentiable on a closed interval $[a,b]$ if f is differentiable on the open interval (a,b) and if the following limits exist:

Right-hand Derivative at a:

$\lim_{h \to 0^+} \dfrac{f(a+h) - f(a)}{h}$

$=$ Slope of tangent at $x = a$

Left-hand Derivative at b:

$\lim_{h \to 0^-} \dfrac{f(b+h) - f(b)}{h}$

$=$ Slope of tangent at $x = b$

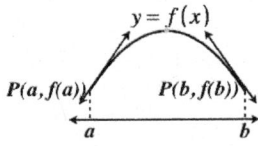

The domain of the derivative f' consists of all numbers at which f is differentiable and also possible endpoints of the domain of f, whenever one-sided limits shown above exist.

Example 4. $f(x) = \sqrt{x}$.   a) sketch the graph of $f(x)$

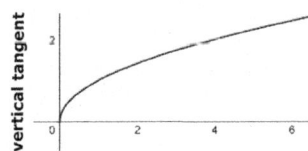

Graph of $f(x)$ has the vertical tangent at $(0,0)$

b) find the domain of $f'(x)$

$f'(x)$ exists for $x > 0$. Check if right-hand derivative exists at $x = 0$

$\lim_{h \to 0^+} \dfrac{f(0+h) - f(0)}{h} = \lim_{h \to 0^+} \dfrac{\sqrt{0+h} - \sqrt{0}}{h} = \lim_{h \to 0^+} \dfrac{\sqrt{h}}{h} = \lim_{h \to 0^+} \dfrac{1}{\sqrt{h}} = \infty$

Right hand derivative DNE at $x = 0 \Rightarrow$ the domain of $f'(x)$ is the set of positive real numbers. $D = \{x \mid x > 0\}$

Example 4 Continued

$f(x) = \sqrt{x}$.   c) find $f'(x)$

$f'(x) = \lim_{h \to 0} \dfrac{\sqrt{x+h} - \sqrt{x}}{h} \cdot \dfrac{\sqrt{x+h} + \sqrt{x}}{\sqrt{x+h} + \sqrt{x}}$

$= \lim_{h \to 0} \dfrac{(x+h) - x}{h(\sqrt{x+h} + \sqrt{x})} = \lim_{h \to 0} \dfrac{1}{\sqrt{x+h} + \sqrt{x}} = \dfrac{1}{2\sqrt{x}}$

$$\boxed{\left(\sqrt{x}\right)' = \dfrac{1}{2\sqrt{x}}}$$

When does $f'(a)$ not exist?

1. Corner, where the one-sided derivatives differ

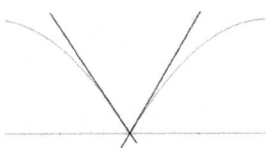 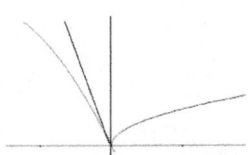

2. Cusp, where the slopes of tangents approach ∞ on one side and $-\infty$ on the other

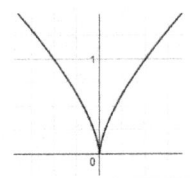

3. Vertical tangent, where the slopes of tangents approach either ∞ or $-\infty$

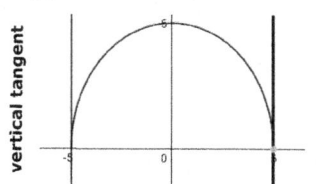 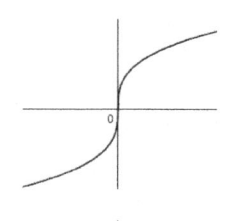

4. Discontinuity, where both or one one-sided derivative is nonexistant

Example 5

Use the graph of $f(x) = \dfrac{1}{x}$ to determine if it is differentiable on the given interval.

$(a) [0, 2]$
No, f is not differentiable at $x = 0$ because f does not exist at $x = 0$.

$(b) [1, 3]$
Yes, because f is smooth and continuous on $[1, 3]$.

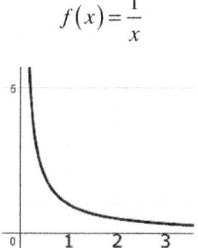

$f(x) = \dfrac{1}{x}$

Example 6

Use the graph of $f(x) = \sqrt{4-x}$ to determine if it is differentiable on the given interval.

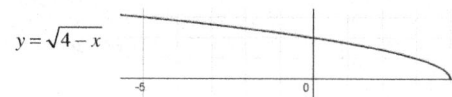

$y = \sqrt{4-x}$

$(a) [0, 4]$ No, the left hand derivative at $x = 4$ does not exist, graph has vertical tangent at $x = 4$.
So f is not differentiable at $x = 4$ and on $[0, 4]$.
$(b) [-5, 0]$ Yes, f is smooth and continuous on $[-5, 0]$ and f' exists (f is differentiable) for every x on $[-5, 0]$.

Theorem

If a function f is differentiable at a, then f is continuous at a.

f is differentiable at $a \Rightarrow f'(x) = \lim\limits_{x \to a} \dfrac{f(x) - f(a)}{x - a}$ exists

The following is a correct statement (left side = right side):

$f(x) = \dfrac{f(x) - f(a)}{x - a} \cdot (x - a) + f(a)$

$\lim\limits_{x \to a} f(x) = \lim\limits_{x \to a} \dfrac{f(x) - f(a)}{x - a} \cdot \lim\limits_{x \to a} (x - a) + \lim\limits_{x \to a} f(a)$

$= f'(a) \cdot 0 + f(a) = f(a)$

$f(x)$ is continuous at $x = a$ if $\lim\limits_{x \to a} f(x)$ exists, $f(a)$ exists, and $\lim\limits_{x \to a} f(x) = f(a)$

f is continuous at a

Use this space for notes.

Use this space for notes.

2.3 Techniques of Differentiation

Formulas for Derivatives

$(I)\; b' = 0$

$(II)\; (mx+b)' = m$

$(III)\; (x^n)' = nx^{n-1}$

$(IV)\; [cf(x)]' = cf'(x)$

$(V)\; [f(x)+g(x)]' = f'(x)+g'(x)$

$(VI)\; [f(x)-g(x)]' = f'(x)-g'(x)$

$f(x)$ & $g(x)$ denote differentiable functions. c, m, and b are real numbers, and n is a rational number.

$(II)\;(mx+b)' = m$ **proof**

$f'(x) = \lim_{h\to 0} \dfrac{f(x+h)-f(x)}{h}$

$f'(x) = \lim_{h\to 0} \dfrac{(m(x+h)+b)-(mx+b)}{h}$

$= \lim_{h\to 0} \dfrac{mx+mh+b-mx-b}{h} = \lim_{h\to 0} \dfrac{mh}{h} = m$

$(I)\; b' = 0$ **proof**

If $m=0$, $(0\cdot x + b)' = 0$

$b' = 0$

Power Rule: $III.\;(x^n)' = nx^{n-1}$ **Proof**

$f'(x) = \lim_{h\to 0} \dfrac{f(x+h)-f(x)}{h} \Rightarrow (x^n)' = \lim_{h\to 0} \dfrac{(x+h)^n - x^n}{h}$

If n is a positive integer, then we can expand by using the binomial theorem, obtaining

$= \lim_{h\to 0} \dfrac{\left[\cancel{x^n} + nx^{n-1}\cancel{h} + \dfrac{n(n-1)}{2!}x^{n-2}h^{\cancel{2}} + \ldots + nxh^{\cancel{n-1}} + h^{\cancel{n}}\right] - \cancel{x^n}}{\cancel{h}}$

$= \lim_{h\to 0}\left[nx^{n-1} + \dfrac{n(n-1)}{2!}x^{n-2}\cancel{h} + \ldots + nxh^{n-2} + h^{n-1}\right] = \boxed{nx^{n-1}}$

Soon we'll prove it when n is negative integer.

$(IV)\;[cf(x)]' = cf'(x)$ **Proof**

$[cf(x)]' = \lim_{h\to 0} \dfrac{cf(x+h)-cf(x)}{h}$

Factor out c inside of the limit

$= \lim_{h\to 0} c\dfrac{f(x+h)-f(x)}{h}$

Move c out of the limit

$= c\lim_{h\to 0} \dfrac{f(x+h)-f(x)}{h}$

$= cf'(x)$

$(V)\;[f(x)+g(x)]' = f'(x)+g'(x)$ **Proof**

$[f(x)+g(x)]' = \lim_{h\to 0} \dfrac{[f(x+h)+g(x+h)]-[f(x)+g(x)]}{h}$

$= \lim_{h\to 0}\left(\dfrac{f(x+h)-f(x)}{h} + \dfrac{g(x+h)-g(x)}{h}\right)$

$= \lim_{h\to 0}\dfrac{f(x+h)-f(x)}{h} + \lim_{h\to 0}\dfrac{g(x+h)-g(x)}{h}$

$= f'(x)+g'(x)$

$(VI)\;[f(x)-g(x)]' = f'(x)-g'(x)$ **Proof**

$(f(x)-g(x))' = (f(x)+(-1)g(x))' =$
$= f'(x)+(-1)g'(x) = f'(x)-g'(x)$

Example 1

Find $f'(x)$ if $f(x) = 2x^4 - 4x^3 + 2x^2 + 5x - 4$.

$$f'(x) = (2x^4)' - (4x^3)' + (2x^2)' + (5x)' - 4'$$

$$\boxed{f'(x) = 8x^3 - 12x^2 + 4x + 5}$$

Product Rule
Proof

$$\boxed{(f(x)g(x))' = f(x)g'(x) + g(x)f'(x)}$$

$$(f(x) \cdot g(x))' = \lim_{h \to 0} \frac{f(x+h)g(x+h) - f(x)g(x)}{h}$$

Add and subtract the same expression

$$= \lim_{h \to 0} \frac{f(x+h)g(x+h) + [-f(x+h)g(x) + f(x+h)g(x)] - f(x)g(x)}{h}$$

$$= \lim_{h \to 0} \left[f(x+h) \cdot \frac{g(x+h) - g(x)}{h} + g(x) \cdot \frac{f(x+h) - f(x)}{h} \right]$$

$$= \lim_{h \to 0} f(x+h) \cdot \lim_{h \to 0} \frac{g(x+h) - g(x)}{h} + \lim_{h \to 0} g(x) \cdot \lim_{h \to 0} \frac{f(x+h) - f(x)}{h}$$

$$\boxed{(f(x)g(x))' = f(x)g'(x) + g(x)f'(x)}$$

Example 2

Find the derivative of $y = (x^3 - 2)(3x^2 + 8x - 4)$

Use Product Rule

$$\frac{dy}{dx} = (x^3 - 2)(3x^2 + 8x - 4)' + (3x^2 + 8x - 4)(x^3 - 2)'$$

$$= (x^3 - 2)(6x + 8) + (3x^2 + 8x - 4)(3x^2)$$

$$= (6x^4 + 8x^3 - 12x - 16) + (9x^4 + 24x^3 - 12x^2)$$

$$\boxed{= 15x^4 + 32x^3 - 12x^2 - 12x - 16}$$

Quotient Rule proof

$$\boxed{\left[\frac{f(x)}{g(x)}\right]' = \frac{g(x)f'(x) - f(x)g'(x)}{[g(x)]^2}}$$

$$\left[\frac{f(x)}{g(x)}\right]' = \lim_{h \to 0} \frac{\frac{f(x+h)}{g(x+h)} - \frac{f(x)}{g(x)}}{h} = \lim_{h \to 0} \frac{g(x)f(x+h) - f(x)g(x+h)}{hg(x+h)g(x)}$$

Add and subtract the same expression

$$= \lim_{h \to 0} \frac{g(x)f(x+h) + [-g(x)f(x) + g(x)f(x)] - f(x)g(x+h)}{hg(x+h)g(x)}$$

$$= \lim_{h \to 0} \frac{g(x)[f(x+h) - f(x)] - f(x)[g(x+h) - g(x)]}{hg(x+h)g(x)}$$

$$= \lim_{h \to 0} \frac{g(x)\left[\frac{f(x+h) - f(x)}{h}\right] - f(x)\left[\frac{g(x+h) - g(x)}{h}\right]}{g(x+h)g(x)}$$

$$= \frac{g(x)f'(x) - f(x)g'(x)}{[g(x)]^2}$$

Example 3

Find the equation of the tangent line to the graph of $y = \dfrac{3x^2 - 2}{4x - 6}$ at the point $\left(1, -\dfrac{1}{2}\right)$.

Equation of the tangent line: $y - y_0 = m(x - x_0)$

Slope of the tangent line at $x = 1$ is: $m = \left.\dfrac{dy}{dx}\right|_{at\ x=1}$

Use Quotient Rule: $\dfrac{dy}{dx} = \dfrac{(3x^2 - 2)'(4x - 6) - (4x - 6)'(3x^2 - 2)}{(4x - 6)^2}$

$\dfrac{dy}{dx} = \dfrac{6x(4x - 6) - 4(3x^2 - 2)}{(4x - 6)^2} = \dfrac{24x^2 - 36x - 12x^2 + 8}{(4x - 6)^2} = \dfrac{12x^2 - 36x + 8}{(4x - 6)^2}$

$\left.\dfrac{dy}{dx}\right|_{at\ x=1} = \dfrac{12 \cdot 1^2 - 36 \cdot 1 + 8}{(4 \cdot 1 - 6)^2} = -4$

Equation of the tangent line is $\boxed{y + \dfrac{1}{2} = -4(x - 1)}$

Reciprocal Rule

$$\boxed{\left(\frac{1}{g(x)}\right)' = -\frac{g'(x)}{(g(x))^2}}$$

Proof

To prove, use the Quotient Rule with $f(x) = 1$

$$\left(\frac{1}{g(x)}\right)' = \frac{1' \cdot g(x) - g'(x) \cdot 1}{(g(x))^2} = \boxed{-\frac{g'(x)}{(g(x))^2}}$$

Example 4

Find the slope of the normal line to the graph of the function
$f(x) = \dfrac{2}{x^2 - 4}$ at the point $x = 4$.

Slope of tangent line of $f(x)$ at $x = 4$ is $f'(x)$

Use the Reciprocal Rule to find $f'(x)$

$f'(x) = \left(\dfrac{2}{x^2-4}\right)' = \dfrac{-2(x^2-4)'}{(x^2-4)^2}$

$= \dfrac{-2 \cdot 2x}{(x^2-4)^2} = \dfrac{-4x}{(x^2-4)^2}$

$f'(4) = \dfrac{-4 \cdot 4}{(4^2-4)^2} = -\dfrac{1}{9}$

Slope of normal line to the graph of $f(x)$: $m_\perp$

$m_\perp = \dfrac{-1}{m} = \dfrac{-1}{\left(\dfrac{-1}{9}\right)}$

Slope of normal line = 9

Power Rule: $III. (x^n)' = nx^{n-1}$ when n is negative integer.

Proof

Let $n = -k$ (where k is positive integer)

by Reciprocal Rule: $\left(\dfrac{1}{g(x)}\right)' = -\dfrac{g'(x)}{(g(x))^2}$

$(x^n)' = (x^{-k})' = \left(\dfrac{1}{x^k}\right)' = -\dfrac{(x^k)'}{(x^k)^2} =$

$-\dfrac{kx^{k-1}}{x^{2k}} = -kx^{k-1-2k} = -kx^{-k-1} = nx^{n-1}$

Later we will prove it when n is a rational number.
It could be proved when n is a real number. Accept it for now.

Example 5

Find the derivative of $y = 5\sqrt[3]{x^2} - \dfrac{3}{\sqrt{x}} + x^{-5}$ at $x = 1$.

$y' = \left(5x^{2/3} - 3x^{-1/2} + x^{-5}\right)'$ **Simplify y**

$= 5 \cdot \dfrac{2}{3} \cdot x^{-1/3} - 3 \cdot \left(-\dfrac{1}{2}\right) \cdot x^{-3/2} + (-5) \cdot x^{-6}$ **Use Power Rule**

$= \dfrac{10}{3x^{1/3}} + \dfrac{3}{2x^{3/2}} - \dfrac{5}{x^6}$

$y'(1) = \dfrac{10}{3 \cdot 1^{1/3}} + \dfrac{3}{2 \cdot 1^{3/2}} - \dfrac{5}{1^6} = \dfrac{10}{3} + \dfrac{3}{2} - 5 = -\dfrac{1}{6}$

Example 6a

Given $f(x) = x^{1/3}(x^2 - 2x + 1)$, find: $(a)\, f'(x)$

Use Product Rule

$(a)\ f'(x) = x^{1/3}(x^2 - 2x + 1)' + (x^2 - 2x + 1)(x^{1/3})'$

$= x^{1/3}(2x - 2) + (x^2 - 2x + 1)\left(\dfrac{1}{3}x^{-2/3}\right)$

$= x^{1/3}(2x - 2) + \dfrac{(x^2 - 2x + 1)}{3x^{2/3}}$

$= \dfrac{3x(2x-2) + (x^2 - 2x + 1)}{3x^{2/3}} = \dfrac{7x^2 - 8x + 1}{3x^{2/3}}$

Example 6b

Given $f(x) = x^{1/3}(x^2 - 2x + 1)$, find:
$(b)\ x$ – coordinates of the points on f at which the tangent line is either horizontal or vertical.

$f'(x)$ is the slope of the tangent line. From (a): $f'(x) = \dfrac{7x^2 - 8x + 1}{3x^{2/3}}$

Tangent is horizontal: slope = 0, or numerator = 0

$7x^2 - 8x + 1 = 0$

$x = \dfrac{8 \pm \sqrt{64 - 28}}{14} = \dfrac{8 \pm \sqrt{36}}{14}$ $\boxed{x = \dfrac{1}{7}, 1}$

Tangent is vertical: slope DNE, or denominator = 0

$3x^{\frac{2}{3}} = 0$ $\boxed{x = 0}$

Use this space for notes.

2.4a Derivatives of Trigonometric Functions

Mathboat.com

Trigonometric Limits

In the Unit circle (Radius=1): $P(\cos x, \sin x)$

$$0 < |\sin x| < |x|$$

By Sandwich Theorem:

$$\boxed{\lim_{x \to 0} \sin x = 0}$$

$$0 < |1 - \cos x| < |x|$$

By Sandwich Theorem:

$$\lim_{x \to 0}(1 - \cos x) = 0$$

$$\boxed{\lim_{x \to 0} \cos x = 1}$$

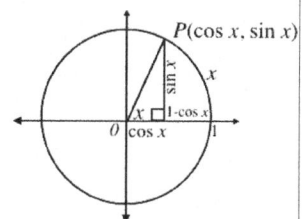

Theorem: $\lim_{x \to 0} \dfrac{\sin x}{x} = 1$ **Proof**

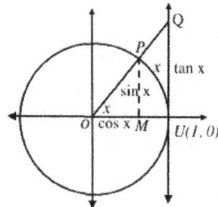

Let $0 < x < \dfrac{\pi}{2}$.

Area of $\triangle OMP = \dfrac{1}{2} \cdot OM \cdot MP = \dfrac{1}{2} \cdot \cos x \sin x$

Area of $\triangle OUQ = \dfrac{1}{2} \cdot 1 \cdot UQ = \dfrac{1}{2} \cdot \tan x$

$UQ = \tan x$ because in $\triangle OUQ$:

$\tan x = \dfrac{\text{opposite}}{\text{adjacent}} = \dfrac{QU}{OU} = \dfrac{QU}{1} = QU$

Area of the sector $= \dfrac{1}{2} r^2 x = \dfrac{1}{2} \cdot 1 \cdot x$

Area of $\triangle OMP <$ Sector Area $<$ Area of $\triangle OUQ$

Theorem: $\lim_{x \to 0} \dfrac{\sin x}{x} = 1$ **Proof continued**

From previous slide:

Area of $\triangle OMP <$ Sector Area $<$ Area of $\triangle OUQ$

$$\dfrac{1}{2}\cos x \sin x < \dfrac{1}{2} x < \dfrac{1}{2} \cdot \dfrac{\sin x}{\cos x}$$

Divide by $\sin x$, multiply by 2:

$$\cos x < \dfrac{x}{\sin x} < \dfrac{1}{\cos x};$$

$$\lim_{x \to 0} \cos x < \lim_{x \to 0} \dfrac{x}{\sin x} < \lim_{x \to 0} \dfrac{1}{\cos x}$$

$$\boxed{\lim_{x \to 0} \cos x = 1 \Rightarrow \lim_{x \to 0} \dfrac{1}{\cos x} = 1}$$

By Sandwich Theorem: $\lim_{x \to 0} \dfrac{\sin x}{x} = 1$

Trigonometric Limits

1. $\boxed{\lim_{x \to 0} \dfrac{\sin x}{x} = 1}$

2. $\boxed{\lim_{x \to 0} \dfrac{1 - \cos x}{x} = 0}$ **Proof**

Let's multiply and divide by the same expression

$$\lim_{x \to 0} \dfrac{1 - \cos x}{x} \cdot \dfrac{1 + \cos x}{1 + \cos x} = \lim_{x \to 0} \dfrac{1 - \cos^2 x}{x(1 + \cos x)}$$

$$= \lim_{x \to 0} \dfrac{\sin^2 x}{x(1 + \cos x)} = \lim_{x \to 0} \dfrac{\sin x}{x} \cdot \lim_{x \to 0} \dfrac{\sin x}{1 + \cos x}$$

$$= 1 \cdot \dfrac{0}{1 + 1} = 0$$

Example 1

$a.\ \lim_{x \to 0} \dfrac{\sin 3x}{x} = 3 \lim_{\substack{x \to 0 \\ 3x \to 0}} \dfrac{\sin 3x}{3x} = 3 \cdot 1 = 3$

$b.\ \lim_{x \to 0} \dfrac{\sin 3x}{\sin 2x} = \lim_{x \to 0} \dfrac{\sin 3x}{\sin 2x} \cdot \dfrac{2x}{3x} \cdot \dfrac{3}{2}$

$= \lim_{\substack{x \to 0 \\ 3x \to 0}} \dfrac{\sin 3x}{3x} \cdot \lim_{\substack{x \to 0 \\ 2x \to 0}} \dfrac{2x}{\sin 2x} \cdot \dfrac{3}{2} = 1 \cdot 1 \cdot \dfrac{3}{2} = \dfrac{3}{2}$

Derivatives of Trigonometric Functions

$(\sin x)' = \cos x$

$(\cos x)' = -\sin x$

$(\tan x)' = \sec^2 x$

$(\cot x)' = -\csc^2 x$

$(\sec x)' = \sec x \tan x$

$(\csc x)' = -\csc x \cot x$

Let's prove them

$(\sin x)' = \cos x$ **Proof.**

Using definition of derivative,

$$(\sin x)' = \lim_{h \to 0} \frac{\sin(x+h) - \sin x}{h}$$

$$= \lim_{h \to 0} \frac{\sin x \cos h + \cos x \sin h - \sin x}{h}$$

$$= \lim_{h \to 0} \frac{\sin x (\cos h - 1) + \cos x \sin h}{h}$$

$$= \lim_{h \to 0} \left(\sin x \left(\frac{\cos h - 1}{h} \right) + \cos x \left(\frac{\sin h}{h} \right) \right)$$

$\lim_{h \to 0} \frac{\cos h - 1}{h} = 0$ $\lim_{h \to 0} \frac{\sin h}{h} = 1$

$$(\sin x)' = (\sin x)(0) + (\cos x)(1) = \cos x$$

$(\cos x)' = -\sin x$ **Proof.**

Using definition of derivative,

$$(\cos x)' = \lim_{h \to 0} \frac{\cos(x+h) - \cos x}{h}$$

$$= \lim_{h \to 0} \frac{\cos x \cos h - \sin x \sin h - \cos x}{h}$$

$$= \lim_{h \to 0} \frac{\cos x (\cos h - 1) - \sin x \sin h}{h}$$

$$= \lim_{h \to 0} \left(\cos x \left(\frac{\cos h - 1}{h} \right) - \sin x \left(\frac{\sin h}{h} \right) \right)$$

$$= (\cos x)(0) - (\sin x)(1) = -\sin x$$

$(\tan x)' = \sec^2 x$ **Proof**

$$(\tan x)' = \left(\frac{\sin x}{\cos x} \right)'$$

Using quotient rule,

$$= \frac{\cos x (\sin x)' - \sin x (\cos x)'}{\cos^2 x}$$

$$= \frac{\cos x (\cos x) - \sin x (-\sin x)}{\cos^2 x}$$

$$= \frac{\cos^2 x + \sin^2 x}{\cos^2 x} = \frac{1}{\cos^2 x} = \sec^2 x$$

$(\sec x)' = \sec x \tan x$ **Proof**

$$(\sec x)' = \left(\frac{1}{\cos x} \right)'$$

Using quotient rule,

$$= \frac{1' \cdot (\cos x) - (\cos x)' \cdot 1}{\cos^2 x}$$

$$= \frac{\sin x}{\cos^2 x}$$

$$= \left(\frac{1}{\cos x} \right) \left(\frac{\sin x}{\cos x} \right) = \sec x \tan x$$

Example 2

Find y' if $y = \dfrac{\sin x}{1 - \cos x}$

Solution

Using quotient rule,

$$y' = \frac{(\sin x)' \cdot (1 - \cos x) - (1 - \cos x)' \cdot (\sin x)}{(1 - \cos x)^2}$$

$$= \frac{(\cos x) \cdot (1 - \cos x) - (0 + \sin x) \cdot (\sin x)}{(1 - \cos x)^2}$$

$$= \frac{\cos x - \cos^2 x - \sin^2 x}{(1 - \cos x)^2} = \frac{\cos x - 1}{(1 - \cos x)^2}$$

$$= -\frac{1 - \cos x}{(1 - \cos x)^2} = -\frac{1}{1 - \cos x}$$

Example 3

Find $g'(x)$ if $g(x) = \sec x \tan x + \cos x$

Solution

Using Product rule,

$g'(x) = (\sec x)(\tan x)' + (\tan x)(\sec x)' + (-\sin x)$

$= (\sec x)(\sec^2 x) + (\tan x)(\sec x \tan x) - \sin x$

$\boxed{= \sec^3 x + \sec x \tan^2 x - \sin x}$

Example 4

(a) Find the slope of the tangent line to the graph of $y = \sin x$ at $x = \dfrac{2\pi}{3}$.

(b) For what values of x is the tangent line horizontal if $0 \leq x \leq 2\pi$?

Solution

(a) The slope of the tangent line at (x, y) is $y' = \cos x$

$y'\big|_{x=\frac{2\pi}{3}} = \cos\dfrac{2\pi}{3} = -\cos\dfrac{\pi}{3} = \boxed{-\dfrac{1}{2}}$

(b) A tangent line is horizontal if its slope is zero.
$y' = 0$; that is, $\cos x = 0$

$x = \boxed{\dfrac{\pi}{2}, \dfrac{3\pi}{2}}$

Example 5

Find an equation of the normal line to the graph of $y = \cot x$ at the point $P\left(\dfrac{\pi}{4}, 1\right)$.

Solution

Equation of a line: $y - y_0 = m(x - x_0)$

First find the slope $m_{\tan}$ of the tangent line:

$y' = (\cot x)' = -\csc^2 x$

$m_{\tan} = -\csc^2\left(\dfrac{\pi}{4}\right) = -\left(\sqrt{2}\right)^2 = -2$

Slope of the normal line is: $m_{\text{normal}} = \dfrac{-1}{m_{\tan}} = \dfrac{1}{2}$

Equation of the normal line is: $\boxed{y - 1 = \dfrac{1}{2}\left(x - \dfrac{\pi}{4}\right)}$

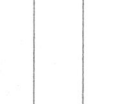

Example 6

Find the first eight derivatives of $f(x) = \sin x$.

$f'(x) = (\sin x)' = \cos x$

$f''(x) = (\cos x)' = -\sin x$

$f'''(x) = (-\sin x)' = -(\sin x)' = -\cos x$

$f^{(4)}(x) = (-\cos x)' = -(\cos x)' = \sin x$

Since $f(x) = \sin x$, it follows that if we continue differentiating, the same pattern repeats, that is:

$f^{(5)}(x) = \cos x$

$f^{(6)}(x) = -\sin x$

$f^{(7)}(x) = -\cos x$

$f^{(8)}(x) = \sin x$

Use this space for notes.

Use this space for notes.

2.4 b Linear Approximation

Let's take a look at the **graph of a function f(x)** and its **tangent line g(x)**.

Near the point of tangency, $x = c$ the tangent line and the function have almost the same graph.
So we can use the tangent line g(x), as an approximation to the function f(x) near $x = c$.
In these cases we call the tangent line the **linear approximation** to the function at $x = c$

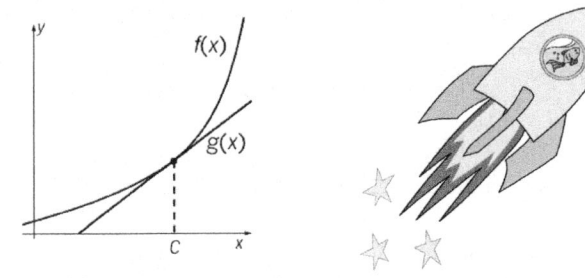

Example 1

a) Find the linear approximation of $f(x) = 1 + \sin x$ at the point $(0,1)$.

The equation of the tangent line of $f(x)$ at the point $(0,1)$:
$y - f(0) = f'(0)(x-0)$

$f'(x) = \cos x$

$f'(0) = \cos 0 = 1$

So, the equation is
$y - 1 = (1)(x - 0)$
$y - 1 = x$
$\boxed{y = x + 1}$

Example 1 Continued

b) Use $g(x) = 1 + x$, the linear approximation to $f(x) = 1 + \sin x$ around $x = 0$, to approximate a value of $f(.75)$.

$f(.75) \approx g(.75) = .75 + 1 = \boxed{1.75}$

At the point $(0,1)$, $g(0)$ is on the curve $f(x)$

At the point $(0.75, 1.75)$, $g(0.75)$ is not on the curve, but is close to the curve, so we can use this as an approximation for $f(0.75)$

Example 2

Find the linear approximation to $f(x) = 2x + 4 + \dfrac{4}{x}$ at $x = -1$ and use it to approximate $f(-0.5)$.

Equation: $y - f(c) = f'(c)(x - c)$

$y - f(-1) = f'(-1)(x - (-1))$

$f'(x) = 2 - \dfrac{4}{x^2}$

$f'(-1) = 2 - \dfrac{4}{(-1)^2} = -2$

Example 2 Continued

$f(-1) = 2(-1) + 4 + \dfrac{4}{-1} = -2$

Plug into formula $\Rightarrow y - (-2) = -2(x - (-1))$

$y + 2 = -2x - 2$

$y = -2x - 4$

Approximate $f(-0.5)$

$f(-0.5) = -2(-0.5) - 4 = \boxed{-3}$

2.5 The Chain Rule

Mathboat.com

The Chain Rule

If $y = f(u), u = g(x)$ and the derivatives $\dfrac{dy}{du}$ and $\dfrac{du}{dx}$ both exist, then the composite function $y = f(g(x))$ has a derivative:

$$\frac{dy}{dx} = \frac{dy}{du} \cdot \frac{du}{dx} = f'(u) \cdot g'(x) = f'(g(x)) \cdot g'(x)$$

$$\boxed{(f(g(x)))' = f'(g(x)) \cdot g'(x)}$$

Derivative of Outer function   TIMES   Derivative of Inner function

---

Illustration that the Chain Rule works.

Find $(\sin 2x)'$ without using the Chain rule.

$(\sin 2x)' = (2\sin x \cos x)' = 2 \cdot (\sin' x \cos x + \cos' x \sin x)$ Use Product Rule

$2 \cdot (\cos^2 x - \sin^2 x) = 2\cos 2x$

Now use the Chain Rule

$$\boxed{(f(g(x)))' = f'(g(x)) \cdot g'(x)}$$

$(\sin 2x)' = \cos 2x \cdot 2 = 2\cos 2x$

Outer function   Inner function

We got the same answers with and without the chain rule!

Example 1

Find $\dfrac{dy}{dx}$ when $y = \sqrt{x^3 + 2}$

Outer function

$$\left(\sqrt{x^3+2}\right)' = \frac{1}{2\sqrt{x^3+2}} \cdot 3x^2 = \boxed{\frac{3x^2}{2\sqrt{x^3+2}}}$$

Inner function

Power Chain Rule

If $y = u^n$ and $u = g(x)$,

then $\boxed{(u^n)' = nu^{n-1} \cdot u'}$

Example 2

Find $\dfrac{dy}{dx}$ if $y = \dfrac{1}{(4x^2 + 6x - 7)^3}$

Inner function   Outer function

$$\left((4x^2+6x-7)^{-3}\right)' = -3(4x^2+6x-7)^{-4} \cdot (8x+6)$$

$$= \boxed{\frac{-3(8x+6)}{(4x^2+6x-7)^4}}$$

Example 3

Find $G'(z)$ if $G(z) = (2z-3)^3 \cdot (3z+4)^4$

Use the Product Rule

$G'(z) = (2z-3)^3 \cdot \left((3z+4)^4\right)' + (3z+4)^4 \cdot \left((2z-3)^3\right)'$

$= (2z-3)^3 \cdot 4(3z+4)^3 \cdot 3 + (3z+4)^4 \cdot 3(2z-3)^2 \cdot 2$

Factor out the common factor

$= 6(2z-3)^2 \cdot (3z+4)^3 \left[2 \cdot (2z-3) + (3z+4)\right]$

$= 6(2z-3)^2 \cdot (3z+4)^3 \cdot (7z-2)$

The Chain Rule For Trigonometric Expressions

If $u = g(x)$ is differentiable function, then

$\boxed{(\sin u)' = \cos u \cdot u'}$ $\boxed{(\cos u)' = -\sin u \cdot u'}$

$\boxed{(\tan u)' = \sec^2 u \cdot u'}$ $\boxed{(\cot u)' = -\csc^2 u \cdot u'}$

$\boxed{(\sec u)' = \sec u \tan u \cdot u'}$ $\boxed{(\csc u)' = -\csc u \cot u \cdot u'}$

Just multiply by *u'* at the end

Example 4

If $y = \cos(6x^4)$, find y'

use the Chain Rule

$\left(\cos(6x^4)\right)' = -\sin(6x^4) \cdot 24x^3$

$\boxed{= -24x^3 \sin(6x^4)}$

Example 5

Find $f'(x)$ if $f(x) = \tan^5 3x$

$\left(\tan^5 3x\right)' = 5\tan^4 3x \cdot \sec^2 3x \cdot 3$

$\boxed{= 15 \tan^4 3x \cdot \sec^2 3x}$

Example 6

Find y' if $y = \sqrt{\sin 4x}$

$\left(\sqrt{\sin 4x}\right)' = \frac{1}{2\sqrt{\sin 4x}} \cdot \cos 4x \cdot 4$

$\boxed{= \frac{2\cos 4x}{\sqrt{\sin 4x}}}$

Example 7

Find the derivative of $f(x) = \left(\cos^2(5x)+1\right)^4$

Solution

$f'(x) = \left(\left(\cos^2(5x)+1\right)^4\right)'$

$= 4\left(\cos^2(5x)+1\right)^3 \cdot 2(\cos(5x)) \cdot (-\sin(5x)) \cdot 5$

$\boxed{= -40\left(\cos^2(5x)+1\right)^3 (\cos(5x))(\sin(5x))}$

2.6. Implicit Differentiation

Implicitly and explicitly defined functions

$$y = 4x^2 + 7$$

y is an explicit function of x since it is written as $y = f(x)$

$$3x^2 - y = 11$$

y is implicit function of x since it is not written as $y = f(x)$

$$4x^2 - 2y^2 - y = 6$$

y is implicitly defined in terms of x

What is the Implicit Differentiation and why do we use it?

If the equation $y^4 + 5y^3 = 5x + 1$ determines an implicit function of x, then there is no obvious way to solve for y in terms of x. The derivative of f may be found by the method of implicit differentiation, in which we differentiate each term of the equation with respect to x.

$$4y^3 \frac{dy}{dx} + 15y^2 \frac{dy}{dx} = 5$$

Why multiply by $\frac{dy}{dx}$?

We must apply a Chain Rule because y is the composite function of x

---

Example 1

Find y' if $5xy^3 - x^2y + 2x^5 + 7x - 3 = 0$ Use Product Rule

Implicitly differentiate

$$5 \cdot \left(x(y^3)' + x'y^3 \right) - \left(x^2 y' + (x^2)' y \right) + 10x^4 + 7 = 0$$

Find each derivative

$$5 \left(x \cdot 3y^2 \frac{dy}{dx} + y^3 \right) - \left(x^2 \frac{dy}{dx} + 2xy \right) + 10x^4 + 7 = 0$$

Move all terms with dy/dx to one side of the equation and factor out dy/dx

$$\frac{dy}{dx}(15xy^2 - x^2) = -5y^3 + 2xy - 10x^4 - 7$$

$$\boxed{\frac{dy}{dx} = \frac{-5y^3 + 2xy - 10x^4 - 7}{15xy^2 - x^2}}$$

Example 2

Find y' if $y = 5x^3 \cos y$ Use product rule

Implicitly differentiate

$$y' = 5\left(x^3 (\cos y)' + (x^3)' \cos y \right)$$

$$\frac{dy}{dx} = 5x^3(-\sin y)\frac{dy}{dx} + 5 \cdot 3x^2 \cos y$$

$$\frac{dy}{dx}(1 + 5x^3 \sin y) = 15x^2 \cos y$$

$$\boxed{\frac{dy}{dx} = \frac{15x^2 \cos y}{1 + 5x^3 \sin y}}$$

Power Rule for the rational powers of x

Proof $\boxed{(x^n)' = nx^{n-1}, \text{where } n = \frac{p}{q}}$

Raise both sides to power q:

$$y = x^{\frac{p}{q}} \Rightarrow y^q = x^p$$

Implicitly differentiate

$$qy^{q-1} \frac{dy}{dx} = px^{p-1}$$

$$\frac{dy}{dx} = \frac{px^{p-1}}{qy^{q-1}}$$

$$\frac{dy}{dx} = \frac{p}{q} \cdot \frac{x^{p-1}}{(x^{p/q})^{q-1}}$$

$$\frac{dy}{dx} = \frac{p}{q} \cdot \frac{x^{p-1}}{x^{p - \frac{p}{q}}}$$

$$\frac{dy}{dx} = \frac{p}{q} x^{(p-1) - \left(p - \frac{p}{q}\right)}$$

$$\frac{dy}{dx} = \frac{p}{q} x^{p - 1 - p + \frac{p}{q}}$$

$$\frac{dy}{dx} = \frac{p}{q} x^{\frac{p}{q} - 1} = nx^{n-1}$$

2.7 Related Rates

Mathboat.com

What are the Related Rates?

Suppose that two variables x and y are functions of variable t:
$x = f(t)$ and $y = g(t)$.

Then $\dfrac{dx}{dt}$ and $\dfrac{dy}{dt}$ are the rates of change of x and y with respect to t.

Sometimes x and y are related by means of an equation, such as:
$x^2 - y^3 + y - 3 = 0$

Differentiate implicitly with respect to t:

$2x\dfrac{dx}{dt} - 3y^2\dfrac{dy}{dt} + \dfrac{dy}{dt} = 0$

The derivatives $\dfrac{dx}{dt}$ and $\dfrac{dy}{dt}$ are called related rates since they related by the means of an equation.

We can find $\dfrac{dx}{dt}$ if $\dfrac{dy}{dt}$ is given or find $\dfrac{dy}{dt}$ if $\dfrac{dx}{dt}$ is given.

Example 1. How fast does the radius of a spherical soap bubble change at the moment when the radius is 1cm if the air is blown into it at the rate of 10 cm³/sec?

Given:
$\dfrac{dV}{dt} = 10\,cm^3/\sec$
$r = 1\,cm$
$\dfrac{dr}{dt} = ?$

$V = \dfrac{4}{3}\pi r^3$

$\dfrac{dV}{dt} = 4\pi r^2 \left(\dfrac{dr}{dt}\right)$

$10 = 4\pi \cdot 1^2 \cdot \dfrac{dr}{dt}$

6) does not apply

$\dfrac{dr}{dt} = \dfrac{10}{4\pi} = \dfrac{5}{2\pi}$

1. Draw Diagram
2. Write down what is given and what is asked for
3. Write the equation that relates the variables.
4. Implicitly differentiate
5. Plug in known values from "Given"
6. Find values that are not directly given and plug them in too.
7. Find the unknown rate of change

Example 2. How fast does the water level drop when a cylindrical tank with a radius of 1 m is drained at the rate of 3 m³/sec?

Given:
$\dfrac{dV}{dt} = -3\,m^3/\sec$
$r = 1\,m$
$\dfrac{dh}{dt} = ?$

$V = \pi r^2 h$
Since r does not change, plug in $r = 1$ m right away
$V = \pi \cdot 1^2 \cdot h$
Implicitly differentiate
$\dfrac{dV}{dt} = \pi 1^2 \dfrac{dh}{dt}$
Plug in known values from "Given"
$-3 = \pi \cdot 1^2 \cdot \dfrac{dh}{dt}$
$\dfrac{dh}{dt} = -\dfrac{3}{\pi}$ (m/sec)

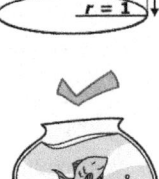

Example 3

A ladder 26 feet long rests on horizontal ground and leans against a vertical wall. The foot of the ladder is pulled away from the wall at the rate of 4 ft/sec. How fast is the top of the ladder sliding down the wall when the foot is 10 ft from the wall?

Given:
$\dfrac{dx}{dt} = 4\,ft/\sec$
$x = 10\,ft$
$L = 26\,ft$
$\dfrac{dy}{dt} = ?$

$x^2 + y^2 = 26^2$
Implicitly differentiate
$2x\dfrac{dx}{dt} + 2y\dfrac{dy}{dt} = 0$

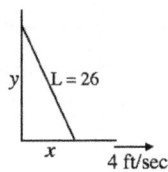

Example 3 Continued

Find values that are not directly given.

When $x = 10$:
$y = \sqrt{26^2 - 10^2} = 24$

Plug in known values

$2 \cdot 10 \cdot 4 + 2 \cdot 24 \cdot \dfrac{dy}{dt} = 0$

$\boxed{\dfrac{dy}{dt} = -\dfrac{2 \cdot 10 \cdot 4}{2 \cdot 24} = -\dfrac{5}{3}\,(ft/\sec)}$

Example 4. Water runs into a conical tank at the rate of 2 ft³/min. The tank stands point down and has a height of 10 ft and a base radius of 5 ft. How fast is the water level rising when the water is 6 ft deep?

$V = \frac{1}{3}\pi x^2 y$

Since we are looking for $\frac{dy}{dt}$ (not $\frac{dx}{dt}$), express x in terms of y.

Using similar triangles,
$\frac{x}{y} = \frac{5}{10} \Rightarrow x = \frac{1}{2}y$

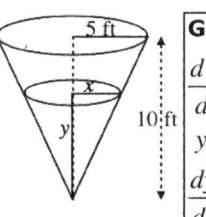

Given:
$\frac{dV}{dt} = 2 ft^3 / min$
$y = 6$ ft
$\frac{dy}{dt} = ?$

$\Rightarrow V = \frac{1}{3}\pi \left(\frac{1}{2}y\right)^2 y \Rightarrow V = \frac{1}{12}\pi y^3$

Example 4 Continued

$V = \frac{1}{12}\pi y^3$

Implicitly differentiate:

$\frac{dV}{dt} = \frac{1}{4}\pi y^2 \cdot \frac{dy}{dt}$

Plug in known values:

$2 = \frac{1}{4}\pi \cdot 6^2 \cdot \frac{dy}{dt}$

$\frac{dy}{dt} = \frac{4 \cdot 2}{36\pi} = \frac{2}{9\pi} \approx 0.071$

Given:
$\frac{dV}{dt} = 2 ft^3 / min$
$y = 6$ ft
$\frac{dy}{dt} = ?$

$\boxed{\frac{dy}{dt} = 0.071 ft / min}$

Example 5. A balloon rising from the ground at 140 ft/min is tracked by a rangefinder at point A, located 500 ft from the point of liftoff. Find the rate at which the **angle** at A and the range **r** are changing when the balloon is 500 ft above the ground.

$\tan\theta = \frac{y}{500}$

Implicitly differentiate

$\sec^2\theta \frac{d\theta}{dt} = \frac{1}{500} \cdot \frac{dy}{dt}$

When $x = y = 500, \theta = \frac{\pi}{4}$

$\sec^2\theta = \left(\sqrt{2}\right)^2 = 2$

Plug in: $2 \cdot \frac{d\theta}{dt} = \frac{1}{500} \cdot 140$

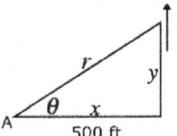

Given:
$\frac{dy}{dt} = 140 ft / min$
$y = 500 ft$
$x = 500 ft$
$d\theta / dt = ?$
$dr / dt = ?$

$\boxed{\frac{d\theta}{dt} = 0.14 rad / min}$

Example 5 Continued

$r^2 = 500^2 + y^2$

Implicitly differentiate

$2r\frac{dr}{dt} = 2y\frac{dy}{dt}$

Find values that are not directly given

When $y = 500$:

$r = \sqrt{500^2 + 500^2} = 500\sqrt{2}$

Plug in known values

$2 \cdot 500\sqrt{2} \cdot \frac{dr}{dt} = 2 \cdot 500 \cdot 140$

$\boxed{\frac{dr}{dt} = 70\sqrt{2} ft / min}$

If in your textbook:

- **Inverse Trig Functions**
- **Derivatives of Inverse Trig Functions,**
- **Derivatives of e^x and a^x,**
- **Derivatives of ln x and $\log_a$ x. Log differentiation**

are introduced **right after Implicit Differentiation or Related Rates** then this is the order of lectures you should follow: 2.6, 2.7, 2.8a, 2.8b, 2.9, 2.10, etc ("Early Transcendental" Approach)

If they are **not** introduced right **after Implicit Differentiation or Related Rates**, then after **2.6 Implicit Differentiation** and **2.7 Related Rates** there should be **3.1** etc. ("Late Transcendental" Approach)

Use this space for notes.

2.8a Inverse Trigonometric Functions

(for "Early Transcendental" approach)
Note: For "Late Transcendental" approach, learn it in chapter 7

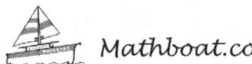 Mathboat.com

Definition

Inverse sine function is $y = \text{Sin}^{-1} x$ if and only if $x = \sin y$ for $[-1, 1]$ as domain and $\left[-\dfrac{\pi}{2}, \dfrac{\pi}{2}\right]$ as range.

Graph of $y = \text{Sin}^{-1} x$ is the image of $y = \sin x$ reflected over $y = x$

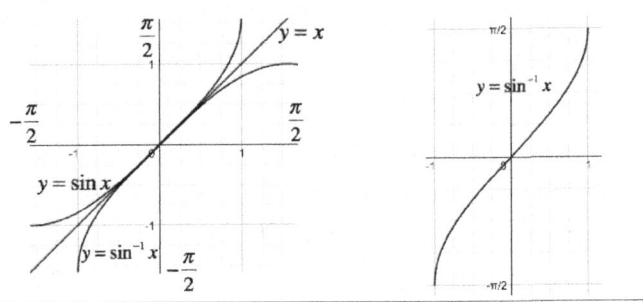

Properties of Arcsin x

1) $\sin\left(\text{Sin}^{-1} x\right) = \sin\left(\text{Arcsin } x\right) = x$ if $-1 \leq x \leq 1$

2) $\text{Sin}^{-1}(\sin x) = \text{Arcsin}(\sin x) = x$ if $-\dfrac{\pi}{2} \leq x \leq \dfrac{\pi}{2}$

x should be inside of the domain of inner function

Examples

1. $\sin\left(\text{Sin}^{-1}\dfrac{1}{3}\right) = \dfrac{1}{3}$, since $-1 < \dfrac{1}{3} < 1$

$\left(\dfrac{1}{3}\text{ is inside of domain of Arcsine}\right)$

Examples Continued

2. $\text{Arcsin}\left(\sin\dfrac{\pi}{3}\right) = \dfrac{\pi}{3}$, since $-\dfrac{\pi}{2} < \dfrac{\pi}{3} < \dfrac{\pi}{2}$

$\left(\dfrac{\pi}{3}\text{ is inside of restricted domain of sine}\right)$

2nd quadrant

3. $\text{Sin}^{-1}\left(\sin\dfrac{3\pi}{4}\right) = \text{Sin}^{-1}\left(\sin\dfrac{\pi}{4}\right) = \dfrac{\pi}{4}$ since $-\dfrac{\pi}{2} < \dfrac{\pi}{4} < \dfrac{\pi}{2}$

$\left(\dfrac{\pi}{4}\text{ is inside of restricted domain of sine}\right)$

Definition

Inverse cosine function is $y = \text{Cos}^{-1} x$ if and only if $x = \cos y$ for $[-1, 1]$ as domain and $[0, \pi]$ as range.

Graph of $y = \text{Cos}^{-1} x$ is the image of $y = \cos x$ reflected over $y = x$.

Since $y = \cos x$ has to pass horizontal test in order to have an inverse function, we need to restrict the domain of $y = \cos x$ from 0 to π.

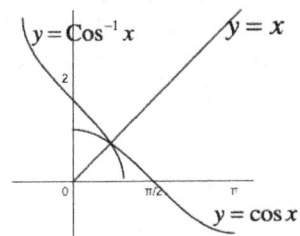 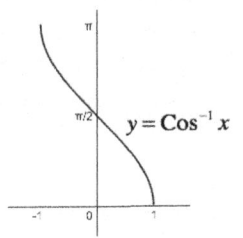

Properties of Arccos x

1) $\cos(\text{Cos}^{-1}x) = \cos(\text{Arccos } x) = x$ if $-1 \leq x \leq 1$
2) $\text{Cos}^{-1}(\cos x) = \text{Arccos}(\cos x) = x$ if $0 \leq x \leq \pi$

x should be inside of the domain of inner function

Examples

1) $\cos\left(\text{Cos}^{-1}\left(-\frac{1}{3}\right)\right) = -\frac{1}{3}$ since $-1 < -\frac{1}{3} < 1$

$\left(-\frac{1}{3}\text{ is inside of domain of Arccosine}\right)$

Examples Continued

2. $\text{Cos}^{-1}\left(\cos\frac{5\pi}{6}\right) = \frac{5\pi}{6}$ since $0 < \frac{5\pi}{6} < \pi$

$\left(\frac{5\pi}{6}\text{ is inside of restricted domain of cosine}\right)$

even function

3. $\text{Cos}^{-1}\left(\cos\left(-\frac{\pi}{6}\right)\right) = \text{Cos}^{-1}\left(\cos\frac{\pi}{6}\right) = \frac{\pi}{6}$ since $0 < \frac{\pi}{6} < \pi$

$\left(\frac{\pi}{6}\text{ is inside of restricted domain of cosine}\right)$

Definition

The inverse tangent function, or arctangent is
$y = \text{Tan}^{-1}x$ if and only if $x = \tan y$
for every x and $\left(-\frac{\pi}{2}, \frac{\pi}{2}\right)$ as range

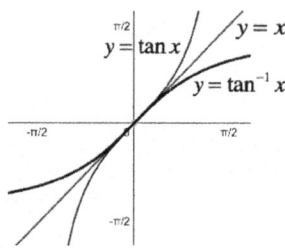
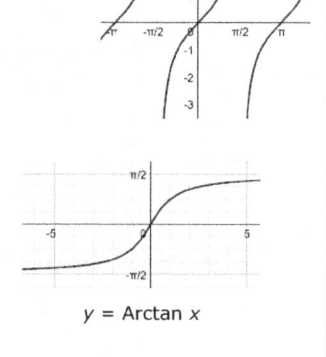

$y = \text{Arctan } x$

Properties of Arctan x

$\tan(\text{Tan}^{-1}x) = \tan(\text{Arctan } x) = x$ for every x

$\text{Tan}^{-1}(\tan x) = \text{Arctan}(\tan x) = x$ if $-\frac{\pi}{2} < x < \frac{\pi}{2}$

x should be inside of the domain of inner function

Examples

1) $\tan\left(\text{Tan}^{-1}\frac{1}{2}\right) = \tan\left(\text{Arctan }\frac{1}{2}\right) = \frac{1}{2}$

Examples of Properties of Arctan x (continued)

2) $\text{Tan}^{-1}\left(\tan\frac{\pi}{3}\right) = \text{Arctan}\left(\tan\frac{\pi}{3}\right) = \frac{\pi}{3}$

3) $\text{Tan}^{-1}\left(\tan\frac{5\pi}{6}\right) = \text{Tan}^{-1}\left(-\tan\frac{\pi}{6}\right) =$

$\text{Tan}^{-1}\left(\tan\left(-\frac{\pi}{6}\right)\right) = -\frac{\pi}{6}$

$\left(\frac{5\pi}{6}\text{ is not inside of restricted domain of tangent}\right)$

Example 1

Find the exact value of $\sec\left(\text{Arctan}\frac{2}{5}\right)$.

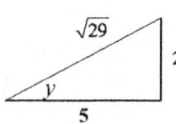

Solution

This is the question we need to answer:

What is secant of an angle, which has tangent $= \frac{2}{5}$?

Let $y = \text{Arctan}\frac{2}{5}$, then $\tan y = \frac{2}{5}$ $\quad \sec y = ?$

By Pythagorean theorem: hypotenuse $= \sqrt{5^2 + 2^2} = \sqrt{29}$

$\boxed{\sec\left(\text{Arctan}\frac{2}{5}\right) = \sec y = \frac{\sqrt{29}}{5}}$

Example 2.

Find the exact value of $\cos\left(\tan^{-1}\left(-\frac{2}{3}\right)\right)$.

This is the question we need to answer:

What is cos of the angle which has a tangent $=-\frac{2}{3}$?

So, $\tan x = -\frac{2}{3}$; Tan is negative in quadrants II and IV.

Restrictions for domain of $\tan x$ are: $\left(-\frac{\pi}{2},\frac{\pi}{2}\right)$ or I and IV.

So, it is Quadrant IV.

cos x is positive in IV.

$$\boxed{\cos x = \frac{3}{\sqrt{13}} = \frac{3\sqrt{13}}{13}}$$

Example 3

Find the exact value of $\tan\left(\text{Arcsin}-\frac{\sqrt{3}}{2}\right)$.

This is the question we need to answer:

What is tangent of the angle which has a sine $=-\frac{\sqrt{3}}{2}$?

sin is negative in III, IV; Restrictions: $\left[-\frac{\pi}{2},\frac{\pi}{2}\right]$

So, it is quadrant IV; $\tan\theta = -\frac{\sqrt{3}}{1} = \boxed{-\sqrt{3}}$

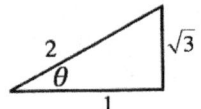

Example 4

Find the exact value of

$$\sin\left(\text{Arctan}\left(-\frac{1}{2}\right) - \text{Arccos}\left(\frac{4}{5}\right)\right).$$

Let $u = \text{Arctan}(-1/2)$ and $v = \text{Arccos}\, 4/5$

$\tan u = -1/2$ and $\cos v = 4/5$

Tangent is negative in II and IV

Restrictions for its domain is $\left(-\frac{\pi}{2},\frac{\pi}{2}\right)$ or: I and IV,

so it is quadrant IV.

Example 4 Continued

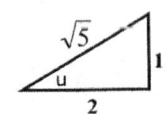 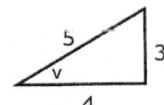

$\sin u = -\frac{1}{\sqrt{5}}$ $\sin v = \frac{3}{5}$

$\cos u = +\frac{2}{\sqrt{5}}$ $\cos v = \frac{4}{5}$

$\sin(u-v) = \sin u \cos v - \cos u \sin v$

$= -\frac{1}{\sqrt{5}}\left(\frac{4}{5}\right) - \frac{2}{\sqrt{5}}\left(\frac{3}{5}\right) = \frac{-4-6}{5\sqrt{5}} = \frac{-2}{\sqrt{5}} = \boxed{\frac{-2\sqrt{5}}{5}}$

Example 5

If $-1 \le x \le 1$, rewrite $\cos(\text{Sin}^{-1} x)$ as an algebraic expression in x.

Solution

Let $y = \text{Sin}^{-1} x$ or, $\sin y = x$

$\cos y = \sqrt{1-\sin^2 y} = \sqrt{1-x^2}$

then: $\cos(\text{Sin}^{-1} x) = \sqrt{1-x^2}$

OR draw the triangle for $\sin y = x$:

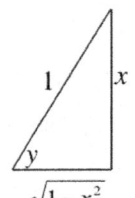

$\cos(\text{Sin}^{-1} x) = \cos y = \frac{\sqrt{1-x^2}}{1} = \boxed{\sqrt{1-x^2}}$

Example 6

Find solutions of $3\sin^2 t + 5\sin t - 2 = 0$ on the interval $[-\pi/2, \pi/2]$

Solution: Equation is quadratic with $x = \sin t$

$3x^2 + 5x - 2 = 0$

$x = \frac{-b \pm \sqrt{b^2-4ac}}{2a} \Rightarrow \sin t = \frac{-5 \pm \sqrt{25+24}}{6} = \frac{-5 \pm 7}{6} = \cancel{-2} \text{ and } \frac{1}{3}$

$\boxed{t = \text{Sin}^{-1}\frac{1}{3} \approx 0.3398} \in [-\pi/2, \pi/2]$

2.8b Derivatives of Inverse Trigonometric Functions

(for "Early Transcendental" approach)

Note: For "Late Transcendental" approach, learn it in chapter 7

Mathboat.com

Derivative of Arcsin x

$$\left(\sin^{-1} x\right)' = \frac{1}{\sqrt{1-x^2}} \quad \text{Proof}$$

$y = \sin^{-1} x \quad -1 < x < 1, \quad -\frac{\pi}{2} < y < \frac{\pi}{2}$

$\sin y = x \quad$ Implicitly differentiate

$\cos y \cdot \dfrac{dy}{dx} = 1$

$\dfrac{dy}{dx} = \dfrac{1}{\cos y} = \dfrac{1}{\sqrt{1-x^2}}$

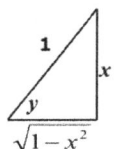

Derivative of Arctan x

$$\boxed{\left(\tan^{-1} x\right)' = \frac{1}{1+x^2}} \quad \text{Proof}$$

$y = \tan^{-1} x, \quad -\dfrac{\pi}{2} < y < \dfrac{\pi}{2}$

$\tan y = x$

Implicitly differentiate

$\sec^2 y \cdot \dfrac{dy}{dx} = 1$

$\dfrac{dy}{dx} = \dfrac{1}{\sec^2 y} = \cos^2 y = \dfrac{1}{1+x^2}$

Derivative of Arcsec x

$$\boxed{\left(\sec^{-1} x\right)' = \frac{1}{x\sqrt{x^2-1}}} \quad \text{Proof}$$

$y = \sec^{-1} x$

$\sec y = x$ for y in either $\left(0, \dfrac{\pi}{2}\right)$ or $\left(\pi, \dfrac{3\pi}{2}\right)$.

Implicitly differentiate.

$\sec y \tan y \cdot \dfrac{dy}{dx} = 1$

$\Rightarrow \dfrac{dy}{dx} = \dfrac{1}{\sec y \tan y}$

$= \dfrac{1}{x\sqrt{x^2-1}}$ for $|x| > 1$

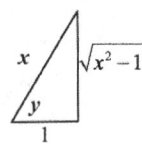

Examples on Derivatives of the Inverse Trig Functions

$$\boxed{\begin{array}{ll} \left(\sin^{-1} u\right)' = \dfrac{1}{\sqrt{1-u^2}} \cdot u' & \left(\tan^{-1} u\right)' = \dfrac{1}{1+u^2} \cdot u' \\ \left(\cos^{-1} u\right)' = -\dfrac{1}{\sqrt{1-u^2}} \cdot u' & \left(\sec^{-1} u\right)' = \dfrac{1}{u\sqrt{u^2-1}} \cdot u' \end{array}}$$

Example 1

$\left(\sin^{-1} 5x\right)' = \dfrac{1}{\sqrt{1-(5x)^2}} \cdot (5x)' = \dfrac{5}{\sqrt{1-25x^2}}$

Example 2

$\left[\arccos(\tan x)\right]' = -\dfrac{1}{\sqrt{1-(\tan x)^2}} (\tan x)' = -\dfrac{\sec^2 x}{\sqrt{1-\tan^2 x}}$

Example 3

$\left(\tan^{-1}(5x^2)\right)' = \dfrac{1}{1+(5x^2)^2} (5x^2)' = \dfrac{10x}{1+25x^4}$

Example 4

$\left[\text{arcsec}(x^5)\right]' = \dfrac{1}{x^5\sqrt{(x^5)^2-1}}(x^5)' = \dfrac{5x^4}{x^5\sqrt{x^{10}-1}} = \dfrac{5}{x\sqrt{x^{10}-1}}$

2.9 Derivatives of e^x and a^x
(for "Early Transcendental" approach)

Note: For "Late Transcendental" approach, learn it in chapter 6

Mathboat.com

The Natural Exponential Function (Review)

The natural exponential function $y = e^x$ is the inverse of the natural logarithmic function $y = \ln x$

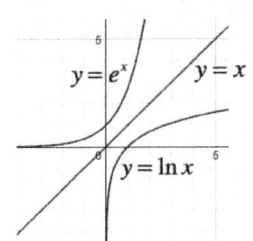

Definition.
If x is any real number,

$$e^x = y \Leftrightarrow \ln y = x$$

$$e \approx 2.71828$$

$\Rightarrow$ irrational number

Theorems (Review)

If p and q are real numbers and r is a rational number, then

$$e^p e^q = e^{p+q}$$

$$\frac{e^p}{e^q} = e^{p-q}$$

$$\left(e^p\right)^q = e^{pq}$$

Theorems (Review)

$$\boxed{\ln e^x = x \text{ for every } x} \qquad \boxed{e^{\ln x} = x \text{ for every } x > 0}$$

Examples:

$$\ln e^{\sqrt{x+2}} = \sqrt{x+2}$$

$$e^{\ln\sqrt{3x+1}} = \sqrt{3x+1}$$

$$e^{5\ln x} = \left(e^{\ln x}\right)^5 = x^5$$

Theorem $\boxed{(e^x)' = e^x}$

Let's evaluate this limit using the calculator:

$$\lim_{h \to 0} \frac{e^h - 1}{h} = ?$$

-0.03	0.985
-0.02	0.990
-0.01	0.995
0	Error
0.01	1.005
0.02	1.010
0.03	1.015

Conclusion:

$$\lim_{h \to 0} \frac{e^h - 1}{h} = 1$$

$\boxed{(e^x)' = e^x}$ proof

By Definition of Derivative,

$$(e^x)' = \lim_{h \to 0} \frac{e^{x+h} - e^x}{h}$$

$$= \lim_{h \to 0} \frac{e^x \cdot e^h - e^x}{h} = \lim_{h \to 0} e^x \cdot \frac{e^h - 1}{h}$$

$$= e^x \cdot \lim_{h \to 0} \frac{e^h - 1}{h} = e^x \cdot 1 = e^x$$

Example 1

Find $f'(x)$, the derivative of $f(x) = x^3 e^{2x}$.

Use Product Rule

$$f'(x) = x^3 \left(e^{2x}\right)' + e^{2x}\left(x^3\right)'$$

Use Chain Rule

$$= x^3 \cdot 2 \cdot e^{2x} + e^{2x} \cdot 3x^2$$

$$\boxed{= x^2 e^{2x}(2x+3)}$$

Example 2

Find $\left(e^{\sqrt{3x^2+2x+4}}\right)'$.

Use Chain Rule

$$\left(e^{\sqrt{3x^2+2x+4}}\right)' = e^{\sqrt{3x^2+2x+4}}\left(\sqrt{3x^2+2x+4}\right)' =$$

$$= e^{\sqrt{3x^2+2x+4}} \cdot \frac{(6x+2)}{2\sqrt{3x^2+2x+4}}$$

$$= \frac{(3x+1)e^{\sqrt{3x^2+2x+4}}}{\sqrt{3x^2+2x+4}}$$

Example 3

If $e^{2x-3xy} = 4$, then $\dfrac{dy}{dx} = ?$

Implicitly differentiate:

$$e^{2x-3xy} \cdot \left(2 - 3(x'y + y'x)\right) = 0$$

$$e^{2x-3xy}\left(2 - 3y - 3x\frac{dy}{dx}\right) = 0$$

Example 3 Continued

$e^{2x-3xy} \neq 0$, always positive

$$2 - 3y - 3x\frac{dy}{dx} = 0$$

$$\frac{dy}{dx} = \boxed{\frac{2-3y}{3x}}$$

Definition

If $f(x) = a^x$, then f is the exponential function with base a

Laws of Exponents:

Let $a > 0$ and $b > 0$. If u and v are any real numbers, then

$$a^u a^v = a^{u+v} \quad \left(a^u\right)^v = a^{uv} \quad (ab)^u = a^u b^u$$

$$\frac{a^u}{a^v} = a^{u-v} \quad \left(\frac{a}{b}\right)^u = \frac{a^u}{b^u}$$

Theorem

$$\boxed{(a^x)' = a^x \ln a}$$

Proof

By Chain Rule, Since $\ln a$ is a constant,

$$(a^x)' = (e^{\ln a^x})' = e^{\ln a^x}(x \ln a)' = a^x \ln a$$

Example 1

Find $f'(x)$ when $f(x) = (x^2+2)^{10} + 10^{x^2+2}$

$$f'(x) = 10(x^2+2)^9(2x) + (10^{x^2+2} \ln 10)(2x)$$

2.10 Derivatives of $y = \ln x$ and $y = \log_a x$. Logarithmic Differentiation.

(for "Early Transcendental" approach)

Note: For "Late Transcendental" approach, learn it in chapter 6

Mathboat.com

Theorem

$(\ln x)' = \dfrac{1}{x}$ if $x > 0$

$y = \ln x \Rightarrow e^y = x$

$\dfrac{d}{dx} e^y = \dfrac{d}{dx} x$

By Chain Rule,

$e^y \dfrac{dy}{dx} = 1$

$\dfrac{dy}{dx} = \dfrac{1}{e^y} = \dfrac{1}{x}$

Theorem

If u is a differentiable function of x and $u > 0$,

$\dfrac{d}{dx} \ln u = \dfrac{1}{u} \dfrac{du}{dx}$

Laws of Logarithms

If $p > 0$ and $q > 0$, then

$\ln pq = \ln p + \ln q$

$\ln \dfrac{p}{q} = \ln p - \ln q$

$\ln p^r = r \ln p$

Example 1

Find the derivative of the function:

$y = \ln(x^2 + 2x + 1)^{\frac{1}{3}}$

Simplify first

$= \dfrac{1}{3} \ln(x+1)^2$

Put exponent 2 in front of ln

$= \dfrac{2}{3} \ln(x+1)$

$y' = \dfrac{2}{3} \cdot \dfrac{1}{x+1} \cdot 1 \quad \boxed{= \dfrac{2}{3(x+1)}}$

Example 2

Find the derivative of the function:

$y = \ln \sqrt[3]{\dfrac{x^2-4}{x^2+4}}$

Simplify first

$y = \ln \left(\dfrac{x^2-4}{x^2+4}\right)^{\frac{1}{3}}$

Put 1/3 in front of ln and use: $\ln\left(\dfrac{p}{q}\right) = \ln p - \ln q$

$y = \dfrac{1}{3}\left[\ln(x^2-4) - \ln(x^2+4)\right]$

By Chain rule,

$\dfrac{dy}{dx} = \dfrac{1}{3}\left(\dfrac{1}{x^2-4} \cdot 2x - \dfrac{1}{x^2+4} \cdot 2x\right)$

$= \dfrac{2x}{3}\left(\dfrac{1}{x^2-4} - \dfrac{1}{x^2+4}\right)$

$= \dfrac{2x}{3}\left(\dfrac{x^2+4 - x^2+4}{(x^2+4)(x^2-4)}\right)$

$= \dfrac{2x}{3}\left(\dfrac{8}{(x^2+4)(x^2-4)}\right)$

$\boxed{\dfrac{dy}{dx} = \dfrac{16x}{3(x^2+4)(x^2-4)}}$

Example 3

Find the derivative of the function defined by equation:

$3y - x^2 + \ln xy = 2$

Use $\ln(pq) = \ln p + \ln q$

$3y - x^2 + \ln x + \ln y = 2$

Implicitly differentiate:

$3\dfrac{dy}{dx} - 2x + \dfrac{1}{x} + \dfrac{1}{y} \cdot \dfrac{dy}{dx} = 0$

Factor out dy/dx

$\dfrac{dy}{dx}\left(3 + \dfrac{1}{y}\right) = 2x - \dfrac{1}{x}$

$\dfrac{dy}{dx} = \dfrac{\dfrac{2x^2-1}{x}}{\dfrac{3y+1}{y}}$

$\boxed{\dfrac{dy}{dx} = \dfrac{(2x^2-1)y}{(3y+1)x}}$

The Definition of $\log_a x$:

$y = \log_a x$ if and only if $x = a^y$

Change of Base Formula: $\log_a x = \dfrac{\ln x}{\ln a}$

Proof:
$$y = \log_a x$$
$$x = a^y$$
$$\ln x = y \ln a$$
$$y = \dfrac{\ln x}{\ln a} \Rightarrow \boxed{\log_a x = \dfrac{\ln x}{\ln a}}$$

Theorem

$$(\log_a x)' = \left(\dfrac{\ln x}{\ln a}\right)' = \dfrac{1}{\ln a} \cdot \dfrac{1}{x}$$

Example 4
If $y = \log(2x+3)$, find $f'(x)$.
$$f'(x) = (\log(2x+3))'$$
$$= \left(\dfrac{\ln(2x+30)}{\ln 10}\right)'$$
$$= \dfrac{1}{\ln 10} \cdot \dfrac{1}{2x+3} \cdot 2 = \dfrac{2}{(2x+3)\ln 10}$$

GUIDELINES FOR LOGARITHMIC DIFFERENTIATION

$y = f(x)$

$\ln y = \ln f(x)$ — Take ln on both sides

$\dfrac{1}{y} \cdot \dfrac{dy}{dx} = (\ln f(x))'$ — Implicitly differentiate

$\dfrac{dy}{dx} = y \cdot (\ln f(x))'$ — Multiply both sides by y

$\dfrac{dy}{dx} = f(x) \cdot (\ln f(x))'$ — Replace y with $f(x)$

Example 5
Find derivative of the function:

$y = \dfrac{(3x-2)^4}{\sqrt{4x+1}}$ ⇒ Take ln on both sides

$\ln y = \ln \dfrac{(3x-2)^4}{\sqrt{4x+1}}$ ⇒ Simplify

$\ln y = \ln(3x-2)^4 - \ln\sqrt{4x+1}$

$\ln y = 4\ln(3x-2) - \dfrac{1}{2}\ln(4x+1)$

Implicitly differentiate

$(\ln y)' = (4\ln(3x-2))' - \left(\dfrac{1}{2}\ln(4x+1)\right)'$

$\dfrac{1}{y}\dfrac{dy}{dx} = 4 \cdot \dfrac{1 \cdot 3}{3x-2} - \dfrac{1}{2} \cdot \dfrac{1 \cdot 4}{4x+1} = \dfrac{42x+16}{(3x-2)(4x+1)}$

Multiply both sides by y
$\dfrac{dy}{dx} = y \cdot \dfrac{42x+16}{(3x-2)(4x+1)}$

Replace y with original function
$\dfrac{dy}{dx} = \dfrac{(3x-2)^4}{\sqrt{4x+1}} \cdot \dfrac{42x+16}{(3x-2)(4x+1)}$

$\dfrac{dy}{dx} = \dfrac{(3x-2)^3(42x+16)}{(4x+1)^{\frac{3}{2}}}$

Example 6

If $y = x^x$ and $x > 0$, find $\dfrac{dy}{dx}$

$y = x^x$ — Use Log differentiation

$\ln y = x \ln x$ — Take ln on both sides

$\dfrac{1}{y}\dfrac{dy}{dx} = x(\ln x)' + \ln x \cdot x'$ — Implicitly differentiate

$\dfrac{dy}{dx} = (1+\ln x)y$ — Multiply both sides by y

$\dfrac{dy}{dx} = (1+\ln x) \cdot x^x$ — Replace y with original function

Example 7

If $y = x^{3x+4}$ and $x > 0$, find $\dfrac{dy}{dx}$

$y = x^{3x+4}$ — Use Log differentiation

$\ln y = \ln x^{3x+4}$ — Take ln on both sides

$(\ln y)' = ((3x+4)\ln x)'$ — Implicitly differentiate

$\dfrac{1}{y} \cdot \dfrac{dy}{dx} = (3x+4)' \ln x + (\ln x)'(3x+4)$ — Use Product Rule on the right and Chain Rule on the left

$\dfrac{1}{y} \cdot \dfrac{dy}{dx} = 3\ln x + \dfrac{1}{x}(3x+4)$ — Find each derivative

$\dfrac{dy}{dx} = \left(3\ln x + 3 + \dfrac{4}{x}\right) \cdot x^{3x+4}$ — Multiply both sides by y, Replace y with original function

3.1 Extrema of functions

Definition of Local Minimum

$f(c)$ is the local minimum value of f if $f(c) \leq f(x)$ for every x in I.

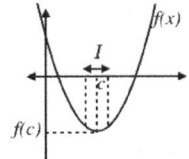
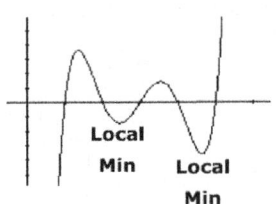

Definition of Local Maximum

$f(c)$ is the local maximum value of f if $f(c) \geq f(x)$ for every x in I.

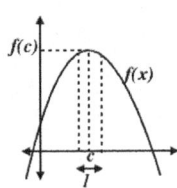
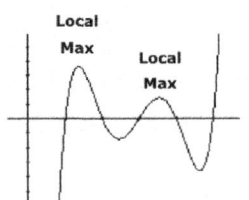

Theorem

If a function f has a local extremum at a number c in an open interval, then either $f'(c)=0$ or $f'(c)$ doesn't exist.

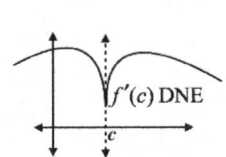
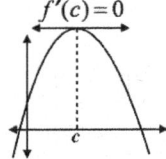

Critical numbers

A number c in the domain of a function f is a critical number of f if either $f'(c)=0$ or $f'(c)$ does not exist.

Example 1

If $f(x) = 4x^3$, prove that f has no local extrema.

Proof Since a local extremum must occur at a critical number, find the critical numbers.

Take the derivative: $f'(x) = 4 \cdot 3x^2$

$f'(x) = 0 \qquad f'(x) \text{ DNE}$
$12x^2 = 0$
$x = 0$

If $x < 0 \to f(x) < 0$, if $x > 0 \to f(x) > 0$

So, $x = 0$ is not the lowest or the highest point on the small interval around it.

The function has no local extrema.

Example 2

Find the critical numbers of $f(x) = \dfrac{2x-3}{x^2-1}$

$f'(x) = \dfrac{(2x-3)'(x^2-1) - (x^2-1)'(2x-3)}{(x^2-1)^2} = \dfrac{2(x^2-1) - 2x(2x-3)}{(x^2-1)^2}$

$= \dfrac{2x^2 - 2 - 4x^2 + 6x}{(x^2-1)^2} = \dfrac{-2x^2 + 6x - 2}{(x^2-1)^2} = \dfrac{-2(x^2 - 3x + 1)}{(x^2-1)^2}$

$f'(x) = 0 \qquad\qquad\text{or}\qquad\qquad f'(x) \text{ DNE}$

$x^2 - 3x + 1 = 0 \qquad\qquad\qquad x^2 - 1 = 0$

$x = \dfrac{3 \pm \sqrt{9-4}}{2} = \dfrac{3 \pm \sqrt{5}}{2} \qquad x = \pm 1$, but they are not in the domain

Critical numbers are $\dfrac{3+\sqrt{5}}{2}, \dfrac{3-\sqrt{5}}{2}$

Example 3

Find the critical numbers of $f(x) = (2x+3)^2 \cdot \sqrt[3]{3x+5}$

$f'(x) = \left((2x+3)^2\right)' \cdot \sqrt[3]{3x+5} + \left(\sqrt[3]{3x+5}\right)'(2x+3)^2$

$f'(x) = \left(2(2x+3) \cdot 2\right) \cdot \sqrt[3]{3x+5} + \left(\frac{1}{3}(3x+5)^{-\frac{2}{3}} \cdot 3\right)(2x+3)^2$

$f'(x) = 4(2x+3) \cdot \sqrt[3]{3x+5} + \frac{(2x+3)^2}{(3x+5)^{\frac{2}{3}}}$

Example 3 Continued

$f'(x) = \frac{4(2x+3) \cdot (3x+5) + (2x+3)^2}{(3x+5)^{\frac{2}{3}}}$

$f'(x) = \frac{(2x+3) \cdot (4(3x+5) + (2x+3))}{(3x+5)^{\frac{2}{3}}}$

$f'(x) = \frac{(2x+3) \cdot (14x+23)}{(3x+5)^{\frac{2}{3}}}$ $f'(x) = 0$ or DNE

3 critical numbers:
$x = -\frac{3}{2}, x = -\frac{23}{14}, x = -\frac{5}{3}$

Example 4

If $f(x) = \cos 2x - 2\sin x$, find the critical numbers of f on the interval $[0, 2\pi]$

Differentiate $f(x)$:

$f'(x) = -2\sin 2x - 2\cos x = 0$ or DNE

$4\sin x \cos x + 2\cos x = 0$

$2\cos x (2\sin x + 1) = 0$

$\cos x = 0$ or $\sin x = -\frac{1}{2}$ (quadr. III, IV)

$x = \frac{\pi}{2}, \frac{3\pi}{2}$ or $x = \frac{7\pi}{6}, \frac{11\pi}{6}$

Absolute Extrema

$f(c)$ is the absolute minimum value of f if $f(c) \leq f(x)$ for every x in the domain of f.

$f(c)$ is the absolute maximum value of f if $f(c) \geq f(x)$ for every x in the domain of f.

To find Absolute Extrema on [a,b]:

1. Find all the critical numbers of f in (a,b)
2. Plug in critical numbers and endpoints of the interval into f
3. The max and the min are the largest and the smallest values of f calculated in #2

Example 5

Find absolute extrema for $f(x) = 3x^2 - 10x + 7$ on $[-1,3]$

1. Find all the critical numbers of f in (a,b)

$f'(x) = 6x - 10 = 0$ or DNE

$6x = 10 \Rightarrow x = \frac{5}{3}$

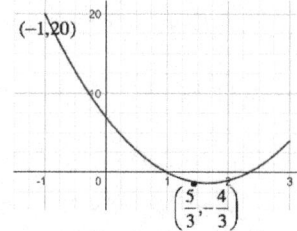

2. Plug in critical numbers and endpoints of the interval into f

$f(-1) = 20$

$f(5/3) = -\frac{4}{3}$

$f(3) = 4$

3. The max and the min are the largest and the smallest values of f calculated in #2.

The absolute max is $f(-1) = 20$

The absolute min is $f(5/3) = -\frac{4}{3}$

Extreme Value Theorem

The Extreme Value Theorem states that if f is a continuous function on the closed interval $[a,b]$, then f attains a maximum value and a minimum value at least once in $[a,b]$.

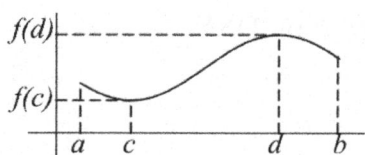

f is continuous on closed interval $[a,b]$. Therefore f has a minimum and maximum value at $f(c)$ and $f(d)$.

A couple Other Cases:

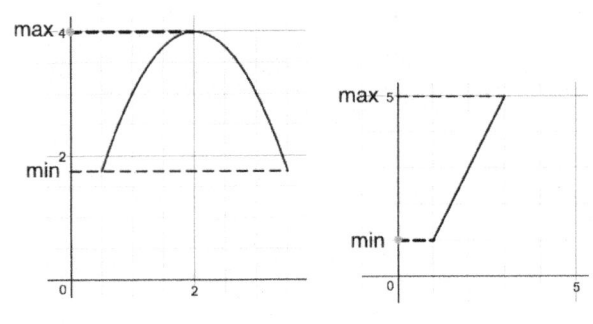

Example 6
What are the maximum and minimum values of $f(x) = x^4 - 3x^3 - 1$ on $[-2,2]$?

f is continuous on $[-2,2]$

$f'(x) = 4x^3 - 9x^2$
$4x^3 - 9x^2 = 0$
$x = 0, \dfrac{9}{4}$

$\dfrac{9}{4}$ is not in the interval $[-2,2]$

$x = 0$ is the only critical point on the interval $[-2,2]$

Test endpoints and critical points:

$f(-2) = (-2)^4 - 3(-2)^3 - 1$
$ = 39 \Leftarrow$ Maximum

$f(0) = (0)^4 - 3(0)^3 - 1$
$ = -1$

$f(2) = (2)^4 - 3(2)^3 - 1$
$ = -9 \Leftarrow$ Minimum

Example 7
For $f(x) = 4 - x^2$, find the absolute extrema of f on the following intervals:

a) Closed interval $[-2,1]$

Max: $f(0) = 4$
Min: $f(-2) = 0$

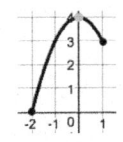

b) Open interval $(-2,1)$

Max: $f(0) = 4$
Min: none

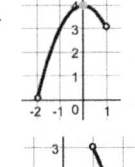

c) $(1,2]$
Max: none
Min: $f(2) = 0$

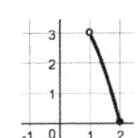

d) $(1,2)$
Max: none
Min: none

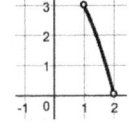

Difference between the Extreme Value of y=f(x) and the Location of the Extreme Value.

- If you are asked to "find a minimum or a maximum of a function", they want **y**-value.

- If you are asked "at which **x** does this function have a min or max", they want **x**-value.

- If you are asked "at which point does this function have a min or max", they want both **x** and **y** coordinates.

Example 8
Find the points of extrema of
$f(x) = 5 - 6x^2 - 2x^3$ on $[-3,1]$.

$f'(x) = -12x - 6x^2$
$f'(x) = 0$ or ~~DNE~~
$f'(x) = -12 - 12x$
$f'(x) = -6x(2 + x)$
$x = 0, x = -2$

Example 8 Continued
Plug in critical numbers and endpoints into f and compare.

$f(0) = 5 - 6(0)^2 - 2(0)^3 = 5$
$f(-2) = 5 - 6(-2)^2 - 2(-2)^3 = -3$
$f(-3) = 5 - 6(-3)^2 - 2(-3)^3 = 5$
$f(1) = 5 - 6(1)^2 - 2(1)^3 = -3$

Maximum at $(0,5), (-3,5)$
Minimum at $(-2,-3), (1,-3)$

3.2 Rolle's Theorem and Mean Value Theorem

Mathboat.com

Rolle's Theorem

Suppose that $y = f(x)$ is continuous on the closed interval $[a,b]$ and differentiable on the open interval (a,b).

If $f(a) = f(b)$,

then there is at least one number c between a and b at which

$f'(c) = 0$.

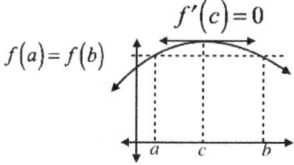

Example 1

Let $f(x) = x^3 - 4x$. Show that f satisfies the hypotheses of Rolle's theorem on the interval $[-2,2]$, and find all real numbers c in the open interval $(-2,2)$, such that $f'(c) = 0$

$f(x) = x^3 - 4x$ is continuous for all x on the interval $[-2,2]$
and differentiable for x on $(-2,2)$

$f(x)$ is a polynomial $\Rightarrow$ continuous and differentiable.

$f(-2) = (-2)^3 - 4 \cdot (-2) = 0$

$f(2) = 2^3 - 4 \cdot 2 = 0$

Hypotheses of Rolle's Theorem are satisfied and $f(-2) = f(2)$

Thus: $f'(c) = 3c^2 - 4 = 0$ at least once between -2 and 2.

$$\boxed{c_1 = \frac{2\sqrt{3}}{3} \text{ and } c_2 = -\frac{2\sqrt{3}}{3}}$$

Mean Value Theorem

If $y = f(x)$ is continuous on the closed interval $[a,b]$ and differentiable on the open interval (a,b), then there is at least one number c between a and b at which

$$\frac{f(b) - f(a)}{b - a} = f'(c)$$

Geometrical Illustration of the Mean Value Theorem

If $y = f(x)$ is continuous on the closed interval $[a,b]$ and differentiable on the open interval (a,b), then there is at least one number c between a and b at which

$$\frac{f(b) - f(a)}{b - a} = f'(c)$$

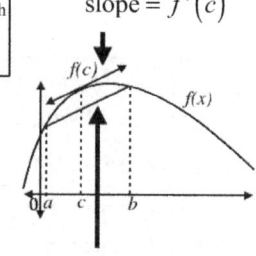

slope $= f'(c)$

slope $= \dfrac{f(b) - f(a)}{b - a}$

$$\frac{f(b) - f(a)}{b - a} = f'(c)$$

Example 2. The graph of $y = f(x)$ on the closed interval $[-4,8]$ is shown in the figure below. If f is continuous on $[-4,8]$ and differentiable on $(-4,8)$, then there exists a c, $-4 < c < 8$, such that

(A) $f'(c) = -4$

(B) $f(c) = -\dfrac{1}{4}$

(C) $f'(c) = \dfrac{1}{4}$

(D) $f'(c) = -\dfrac{1}{4}$

(E) $f(c) = 0$

By the Mean Value Theorem,

$f'(c) = -\dfrac{1}{4}$

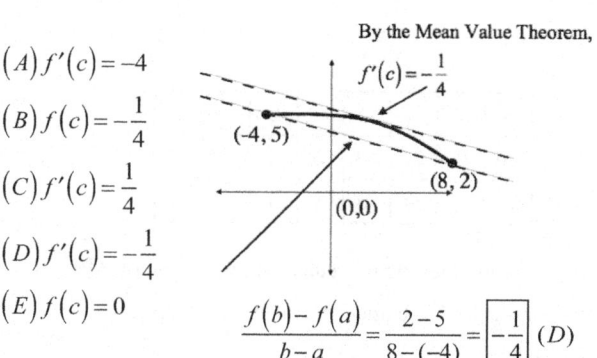

$\dfrac{f(b) - f(a)}{b - a} = \dfrac{2 - 5}{8 - (-4)} = \boxed{-\dfrac{1}{4}}$ (D)

Example 3

If $f(x) = \frac{1}{3}x^2 + 2$, show that f satifies the hypotheses of the Mean Value Theorem on the interval $[-1, 3]$, and find a number c in $(-1, 3)$ that satisfies the conclusion of the theorem.

Solution

$f(x)$ is a Polynomial $\Rightarrow$

f is continuous on $[-1, 3]$,

f is differentiable on $(-1, 3)$.

$$\frac{f(3) - f(-1)}{3 - (-1)} = f'(c)$$

Example 3 Continued

$f'(x) = \frac{2}{3}x,$

$f(3) = 5, f(-1) = 2\frac{1}{3}$

$$\frac{5 - 2\frac{1}{3}}{4} = \frac{2}{3}c$$

$$\frac{\frac{8}{3}}{4} = \frac{2}{3}c$$

$$\frac{2}{3} = \frac{2}{3}c$$

$\boxed{c = 1}$ where $-1 < 1 < 3$

Example 4

If $f(x) = x^3 - 4x + 3$, show that f satisfies the hypotheses of the Mean Value Theorem on the interval $[1, 5]$, and find a number c in the open interval $(1, 5)$ that satisfies the conclusion of the theorem.

Solution

$f(x)$ is a Polynomial, so

f is continuous on $[1, 5]$,

f is differentiable on $(1, 5)$.

$$\frac{f(5) - f(1)}{5 - 1} = f'(c)$$

$f(5) = 108, f(1) = 0$

$f'(x) = 3x^2 - 4$

Example 4 Continued

$$\frac{108 - 0}{4} = 3c^2 - 4$$

$$27 = 3c^2 - 4$$

$$c^2 = \frac{31}{3} \Rightarrow c = \pm \frac{\sqrt{93}}{3}$$

Since $-\frac{\sqrt{93}}{3}$ is not in the interval $(1, 5)$, the answer is $\boxed{\frac{\sqrt{93}}{3}}$

Example 5

Let $f(x)$ be a differentiable function defined on the interval $-5 \leq x \leq 5$. The table below gives the value of $f(x)$ and its derivative $f'(x)$ at several points of the domain.

x	-5	-4	-2	0	2	4	5
f(x)	45	25	13	3	2	4	5
f'(x)	-5	-4	-3	-1	0	1	0

At what point does the line tangent to the graph of $f(x)$ and parallel to the segment between the endpoints intersects the x-axis?

Example 5 Continued   Solution

$f(x)$ is differentiable function, so it continuous. By MVT there is at least one point at which the line tangent to the graph of $f(x)$ is parallel to the segment between the endpoints.

$$\text{slope} = \frac{5 - 45}{5 - (-5)} = \frac{-40}{10} = -4.$$

From the table: the corresponding point is $(-4, 25)$

Equation of tangent line: $y - y_0 = m(x - x_0)$

$y - 25 = -4[x - (-4)] \Rightarrow y = -4x + 9$

To find x-intercept, set $y = 0$

$0 = -4x + 9 \Rightarrow x = \frac{9}{4}$

$\boxed{\left(\frac{9}{4}, 0\right)}$

3.3 The First Derivative Test. Using First Derivative in Graphing

Mathboat.com

Increasing function

A function f **is increasing** on an interval I if $f(p) < f(q)$ for all p and q in I with $p < q$

When $f(x)$ is increasing $f'(x)$ is positive

Decreasing function

A function f **is decreasing** on an interval I if $f(p) > f(q)$ for all p and q in I with $p < q$

When $f(x)$ is decreasing $f'(x)$ is negative

Examine the following graph:

When $f(x)$ is increasing $f'(x)$ is positive

$f'(x) = 0$, critical point

When $f(x)$ is decreasing $f'(x)$ is negative

Example 1

Let $g(x) = x^3 - 3x^2 - 9x + 22$

a) Find the intervals on which g is increasing and decreasing.

$g'(x) = 3x^2 - 6x - 9$

$3(x+1)(x-3) = 0$

$x = -1, 3$ are critical numbers

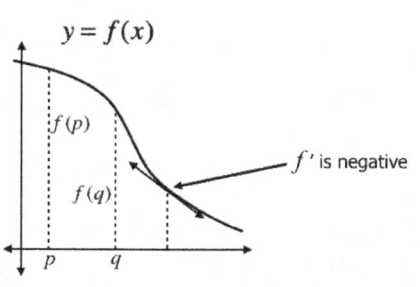

g(x) decreases on: [−1, 3]

g(x) increases on: (−∞, −1] & [3, ∞)

Example 1 Continued

Let $g(x) = x^3 - 3x^2 - 9x + 22$

b) Sketch this graph

$g(-1) = 27$ and $g(3) = -5$

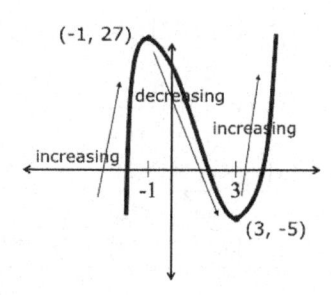

(-1, 27)

(3, -5)

The First-Derivative test

Let f be continuous at c and differentiable on the open interval containing c except possibly at c itself

Local minimum occurs at $x = c$, where $f'(x)$ changes from negative to positive.

Local maximum occurs at $x = c$, where $f'(x)$ changes from positive to negative.

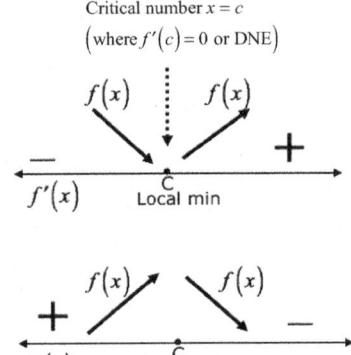

What happens when $f'(c) = 0$, but $f'(x)$ does not change sign around $x = c$?

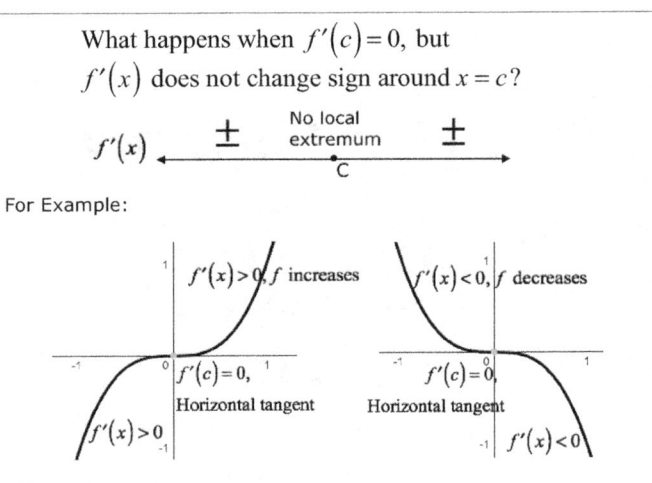

For Example:

No local extremum at x=c

Example 2

Let $\varphi(x) = x^4 - 4x^3 + 12$

a) Find and classify the critical numbers of φ

$\varphi'(x) = 4x^3 - 12x^2$

$\varphi'(x) = 4x^2(x-3) = 0$

$x = 0$ and $x = 3$ are critical numbers

$\varphi'(x)$: $-$ at 0, $-$ at 3 (Local min), $+$

Local minimum occurs at $x = 3$, where $f'(x)$ changes from negative to positive.

No extremum at $x = 0$ and local (or relative) minimum at $x = 3$

Example 2 Continued

Let $\varphi(x) = x^4 - 4x^3 + 12$

b) Graph φ

$\varphi(0) = 12$ and $\varphi(3) = -15$

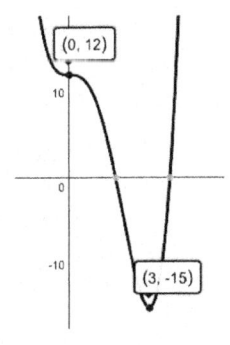

Notice that $\varphi(3) = -15$ is the lowest value of φ for all x, not only near $x = 3$. So, the absolute minimum value of $\varphi(x)$ is -15 at $x = 3$. There is no absolute maximum.

Example 3.
Find the points of local extrema of f and the intervals on which f is increasing and decreasing, and sketch the graph of f.

$f(x) = x^{2/3}(x-7)^2 + 2$

$f'(x) = x^{2/3}\left((x-7)^2\right)' + \left(x^{2/3}\right)'(x-7)^2$

$= x^{2/3}(2(x-7)) + \left(\frac{2}{3}x^{-1/3}\right)(x-7)^2$

$= 2x^{2/3}(x-7) + \frac{2(x-7)^2}{3x^{1/3}}$

$= \frac{6x(x-7) + 2(x-7)^2}{3x^{1/3}}$

$= \frac{(x-7)(6x + 2x - 14)}{3x^{1/3}} = \frac{2(x-7)(4x-7)}{3x^{1/3}}$

$f'(x) = 0$ or DNE

Critical numbers: $x = 7, \frac{7}{4}, 0$

Example 3 Continued

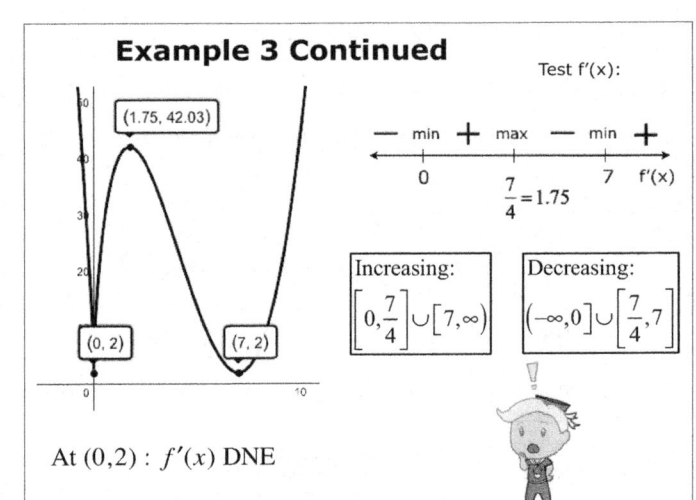

Test $f'(x)$:

$-$ min $+$ max $-$ min $+$ at 0, $\frac{7}{4} = 1.75$, 7 $f'(x)$

Increasing: $\left[0, \frac{7}{4}\right] \cup [7, \infty)$

Decreasing: $(-\infty, 0] \cup \left[\frac{7}{4}, 7\right]$

At $(0, 2)$: $f'(x)$ DNE

3.4 Concavity and the Second Derivative Test

Let's look at the example of how the graphs $f(x), f'(x)$ and $f''(x)$ are related

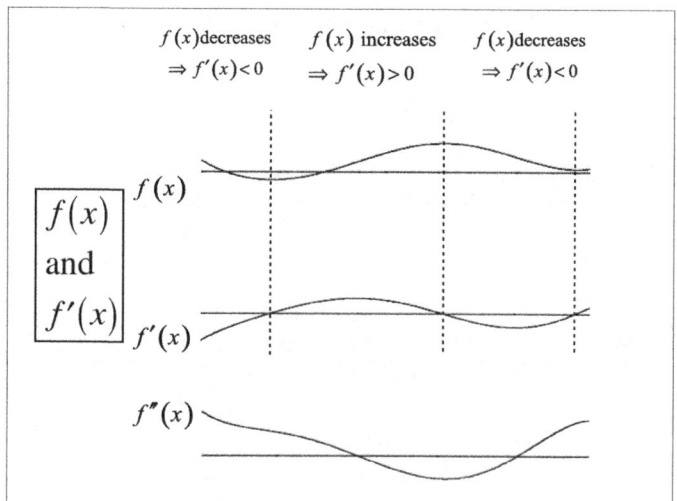

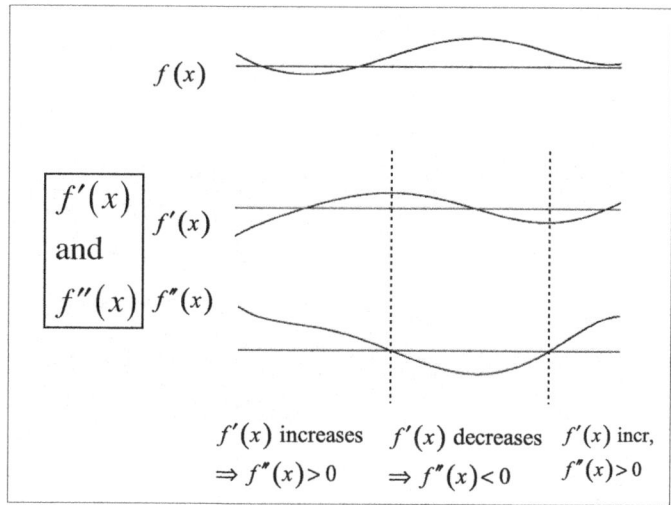

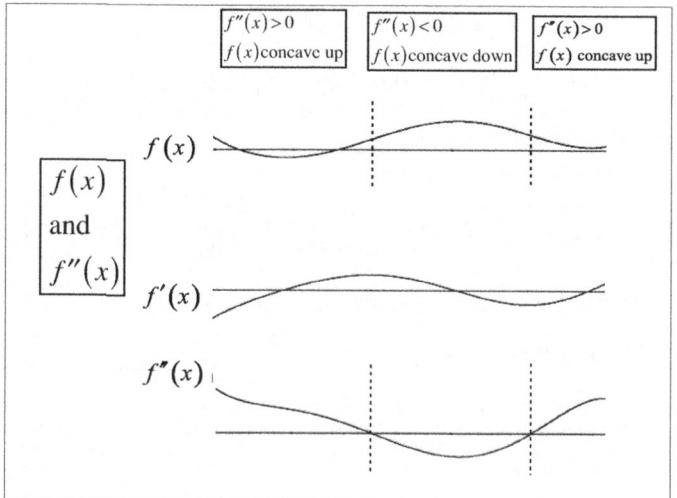

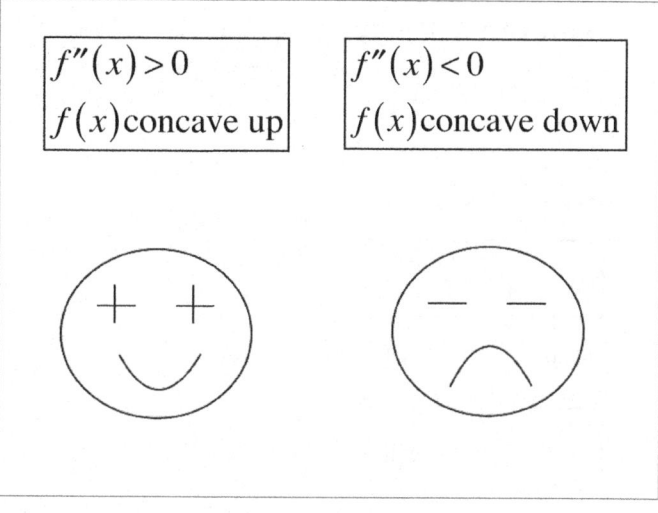

A Point of Inflection occurs at the point where the concavity changes.

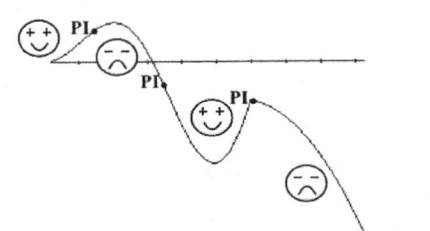

For inflection points, set $f''(x) = 0$ or DNE. If $f''(x)$ changes sign passing through the point, it is a PI.

Example 1

If $f(x) = x^3 + 3x^2 - 2x - 6$, determine intervals on which the graph of $f(x)$ is concave upward or is concave downward and sketch the graph.

$f'(x) = 3x^2 + 6x - 2$
$f''(x) = 6x + 6 = 6(x+1)$
$6(x+1) = 0$ or DNE
$x = -1$
$f(-1) = -1 + 3 + 2 - 6 = -2$

$f''(x)$ — +
         $\odot$ -1 $\smile$

Concave down: $(-\infty, -1)$
Concave up: $(-1, \infty)$

Example 2

The graph of $f'(x)$ is shown below. On which of the following intervals is the graph of $f(x)$ concave up?

a) (-2,4) b) (2,4) c) (-2,2) d) (0,2) e) (-2,0)

Solution

$f(x)$ concave up
→ $f''(x) = [f'(x)]' > 0$
It means $f'(x)$ is increasing.
$f'(x)$ is increasing on [2,4]

$f(x)$ is concave up on (2,4)

2nd Derivative Test for Local Max and Min

Suppose that f is differentiable on an open interval containing c and that $f'(c) = 0$

If $f''(c) > 0$, f has a minimum at c

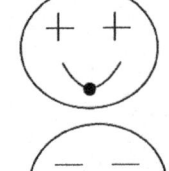

If $f''(c) < 0$, f has a maximum at c

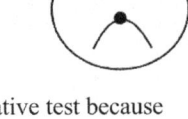

If $f'(c)$ DNE, we don't use 2nd derivative test because $f''(c)$ would not exist and there will be no conclusion

Example 3

If $f(x) = x^{2/3}(x+4)$, find the local extrema, discuss concavity, and find points of inflection.

$f(x) = x^{5/3} + 4x^{2/3}$

$f'(x) = \frac{5}{3}x^{2/3} + \frac{8}{3}x^{-1/3} = \frac{5x^{2/3}}{3} + \frac{8}{3x^{1/3}} + = \frac{5x+8}{3x^{1/3}}$

$f'(x) = \frac{1}{3}\left(\frac{5x+8}{x^{1/3}}\right)$

$f''(x) = \frac{10}{9}x^{-1/3} + \frac{-8}{9}x^{-4/3} = \frac{2}{9}(5x^{-1/3} - 4x^{-4/3}) = \frac{2}{9}\left(\frac{5}{x^{1/3}} - \frac{4}{x^{4/3}}\right)$

$f''(x) = \frac{2}{9}\left(\frac{5x-4}{x^{4/3}}\right)$

Example 3 Continued

$f'(x) = \frac{1}{3}\left(\frac{5x+8}{x^{1/3}}\right) = 0$ or DNE

$f''(x) = \frac{2}{9}\left(\frac{5x-4}{x^{4/3}}\right)$

The critical numbers are x = -8/5, 0.
Use second derivative test for x = -8/5

$f''\left(-\frac{8}{5}\right) < 0$ $\frown$ max $f\left(-\frac{8}{5}\right) \approx 3.28$ Local max $\left(-\frac{8}{5}, 3.28\right)$

If $f'(0)$ DNE, we do not use 2nd derivative test for $x = 0$
Use 1st derivative test.

$f'(x)$ — +
       $-\frac{8}{5}$ 0

Local min at $x = 0$
$f(0) = 0$,
f is not differentiable at $(0,0)$

Example 3 Continued

$$f''(x) = \frac{2}{9}\left(\frac{5x-4}{x^{4/3}}\right)$$ Local max $\left(-\frac{8}{5}, 3.28\right)$ Local min at $x = 0$ $f(0) = 0$, f is not differentiable at $(0,0)$

Inflection points could be the points where $f''(x) = 0$ or DNE. So, check $x = 0, \frac{4}{5}$

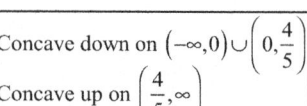

$\left(\frac{4}{5}, 4.14\right)$ is PI

Concave down on $(-\infty, 0) \cup \left(0, \frac{4}{5}\right)$
Concave up on $\left(\frac{4}{5}, \infty\right)$

Example 3 Graph

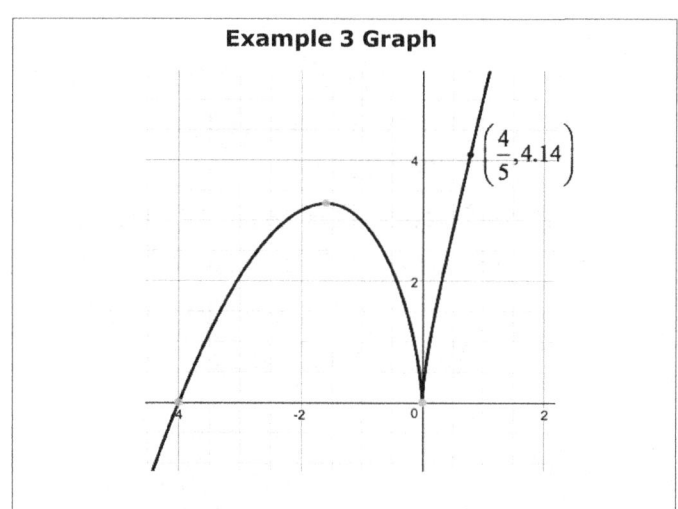

How to Sketch a Graph Based on Information of $f'(x)$ and $f''(x)$

Example 4

Given: $f(0) = 4; f(2) = 2; f(5) = 6.5$

$f'(0) = f'(2) = 0$

$f'(x) > 0$ if $|x-1| > 1$

$f'(x) < 0$ if $|x-1| < 1$

$f''(x) < 0$ if $x < 1$ or if $|x-4| < 1$

$f''(x) > 0$ if $|x-2| < 1$ or if $x > 5$

Example 4 Continued

Solution

$f'(x) > 0$ if $|x-1| > 1$

f increases: $x - 1 > 1$ or $x - 1 < -1$

$x > 2$ or $x < 0$

$f'(x) < 0$ if $|x-1| < 1$

f decreases: $-1 < x - 1 < 1$

$0 < x < 2$

☹ $f''(x) < 0$ if $x < 1$ or if $|x-4| < 1$

$-1 < x - 4 < 1$

$3 < x < 5$

☺ $f''(x) > 0$ if $|x-2| < 1$ or if $x > 5$

$-1 < x - 2 < 1$

$1 < x < 3$

Example 4 Continued

Plot the pts: $f(0) = 4; f(2) = 2; f(5) = 6.5$

$f'(0) = f'(2) = 0 \Rightarrow$ Horizontal tangents at $x = 0, 2$.

f increases: $x > 2$ or $x < 0$

f decreases: $0 < x < 2$

f concaves down if $x < 1$ or $3 < x < 5$

f concaves up if $1 < x < 3$ or $x > 5$

Example 4 Continued

Graph:

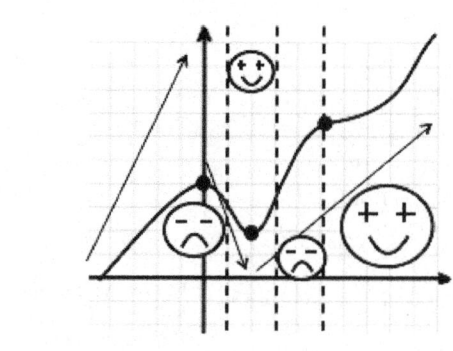

3.5 Summary of Graphical Methods.
Guidelines for sketching the Graph of a Function using First and Second Derivatives.

Mathboat.com

Guidelines for sketching a graph of y = f(x)

There are seven guidelines to follow while sketching a graph.

Example 1. Discuss and sketch the graph of $f(x) = \dfrac{3x^2}{16-x^2}$

1. Domain – find all x that are possible.

   The denominator of a function cannot equal 0.
   $$16 - x^2 \neq 0$$
   $$x^2 \neq 16$$
   $$x \neq -4, 4$$
   The domain of f consists of all real numbers except −4 and 4.
   $$D = \{x | x \neq -4, 4\}$$

2. Continuity – determine whether f is continuous on its domain.

   f is a rational function, it is continuous on its domain, but f has infinite discontinuities at −4 and 4 which are not on the domain

---

3. Intercepts: find the x-and y-intercepts.

y-intercepts: $x = 0$ or find $f(0)$

$$y = f(0) = \frac{3(0)^2}{16-(0)^2} = 0 \quad \boxed{(0,0)}$$

x-intercepts: $y = 0$ or solve $f(x) = 0$

$$f(x) = \frac{3x^2}{16-x^2} = 0$$
$$x^2 = 0 \Rightarrow x = 0 \quad \boxed{(0,0)}$$

The graph intersects both axes at the origin.

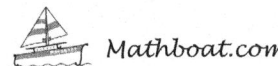
$f(x) = \dfrac{3x^2}{16-x^2}$

4. Symmetry – check to see if the function is odd or even.

If the function is even, the graph will be symmetrical with respect to the y-axis.

If the function is odd, the graph will be symmetrical with respect to the origin.

$$f(-x) = \frac{3(-x)^2}{16-(-x)^2} = \frac{3x^2}{16-x^2} = f(x)$$

Since $f(-x) = f(x)$, this function is even and is symmetrical with respect to the y-axis.

$f(x) = \dfrac{3x^2}{16-x^2}$

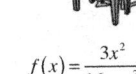

---

5 a. Critical Numbers – find values of x on the domain of $f(x)$ where $f'(x) = 0$ or $f'(x)$ DNE

$$f'(x) = \frac{(3x^2)'(16-x^2) - (16-x^2)'(3x^2)}{(16-x^2)^2}$$
$$= \frac{(6x)(16-x^2) - (-2x)(3x^2)}{(16-x^2)^2}$$
$$= \frac{(96x - 6x^3) - (-6x^3)}{(16-x^2)^2} = \frac{96x}{(16-x^2)^2}$$

$f'(x) = 0$ or $f'(x)$ DNE

$$\frac{96x}{(16-x^2)^2} = 0 \quad \text{or} \quad (16-x^2)^2 = 0$$

$\boxed{x = 0, \text{ critical number}}$ $x = -4$ and $4 \Rightarrow$ Not on the domain. Not critical numbers. But increasing/decreasing could change around them.

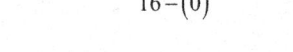

5 b. Local Extrema
Use First Derivative test to find local extrema.

Increasing/decreasing
Increasing if $f'(x) > 0$, decreasing if $f'(x) < 0$

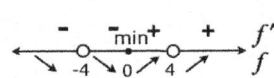

Increasing: $[0, 4) \cup (4, \infty)$
Decreasing: $(-\infty, -4) \cup (-4, 0]$

Local min at $x = 0 : (0,0)$

$f(x) = \dfrac{3x^2}{16-x^2}$

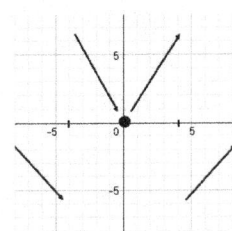

6 a. Points of Inflection

Find $f''(x)$, set $f''(x) = 0$ and $f''(x) = DNE$,
Find x where concavity of $f(x)$ changes

$$f''(x) = \left[\frac{96x}{(16-x^2)^2}\right]'$$

$$= 96 \cdot \frac{x' \cdot (16-x^2)^2 - \left((16-x^2)^2\right)' \cdot x}{(16-x^2)^4}$$

$$= 96 \cdot \frac{(16-x^2)^2 - 2(16-x^2) \cdot (-2x) \cdot x}{(16-x^2)^4}$$

$$= 96 \cdot \frac{(16-x^2)\left((16-x^2) + 4x^2\right)}{(16-x^2)^{4\,3}} = \frac{96(16+3x^2)}{(16-x^2)^3}$$

$f''(x) \neq 0$ or $f''(x)$ DNE
$(16-x^2)^3 = 0$
$x = -4, 4$

These are not PIs, not on the domain, but concavity could change around them.

No points of inflection.

6 b. Concavity

If $f''(x) > 0$, graph is concave up
If $f''(x) < 0$, graph is concave down.

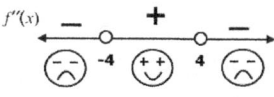

Concave down on $(-\infty, -4) \cup (4, \infty)$

Concave up on $(-4, 4)$

$$f(x) = \frac{3x^2}{16-x^2}$$

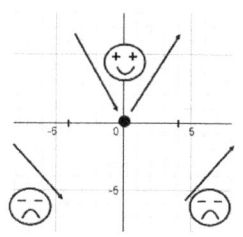

7. Asymptotes: lines graph is approaching to.

7a. Horizontal asymptote: if $\lim_{x \to \infty} f(x) = L$
or if $\lim_{x \to -\infty} f(x) = L$, then the line $y = L$
is a horizontal asymptote.

$$\lim_{x \to +\infty} f(x) \qquad \lim_{x \to -\infty} f(x)$$

$$= \lim_{x \to +\infty} \frac{3x^2}{16-x^2} \quad = \lim_{x \to -\infty} \frac{3x^2}{16-x^2}$$

$$= \lim_{x \to +\infty} \frac{3}{\frac{16}{x^2}-1} = -3 \quad = \lim_{x \to -\infty} \frac{3}{\frac{16}{x^2}-1} = -3$$

Horizontal asymptote: $y = -3$

Or: Power of numerator = power of denominator
so Horizontal asymptote: $y = \frac{3}{-1} = -3$

$$f(x) = \frac{3x^2}{16-x^2}$$

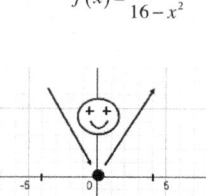

7b. Vertical asymptote: If $\lim_{x \to a^+} f(x)$ or $\lim_{x \to a^-} f(x)$ is equal to either $+\infty$ or $-\infty$, then the line $x = a$ is a vertical asymptote.
Or, the vertical asymptotes correspond to the zero of the denominator $16 - x^2$

$$16 - x^2 = 0$$
$$x = 4, \ x = -4$$

Vertical asymptotes are $x = -4$ and $x = 4$

$$f(x) = \frac{3x^2}{16-x^2}$$

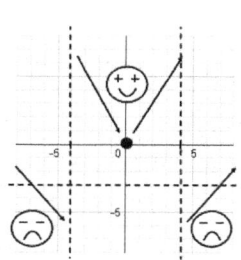

Time to graph!

$$f(x) = \frac{3x^2}{16-x^2}$$

Check using graphing calculator:

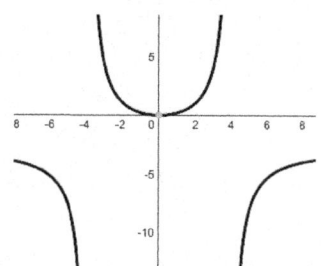
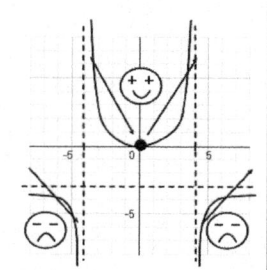

Example 2. Discuss and sketch the graph of $f(x) = \frac{x^2-16}{2x-6}$

1. Domain: $2x - 6 \neq 0$
   $x \neq 3$
   Domain = all real numbers, except $x = 3$

   $D = \{x | x \neq 3\}$

2. Continuity
   Continuous on its domain, But it has infinite discontinuity at $x=3$

3. Intercepts:
   x-intercepts: $y = 0 \quad x^2 - 16 = 0$
   $x = -4, 4$
   y-intercepts: $x = 0$
   $y = \frac{0^2 - 16}{2(0) - 6} \quad y = \frac{8}{3}$

4. Symmetry:
   $f(-x) \neq f(x)$ or $-f(x)$
   $f(x)$ is neither odd nor even.
   So $f(x)$ is not symmetric with respect to the y-axis or origin.

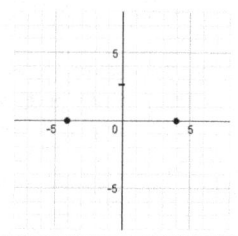

5. Critical Numbers and Extrema:

$$f(x) = \frac{x^2 - 16}{2x - 6}$$

$$f'(x) = \frac{(x^2 - 16)'(2x-6) - (2x-6)'(x^2-16)}{(2x-6)^2}$$

$$= \frac{2x(2x-6) - 2(x^2-16)}{(2x-6)^2} = \frac{2x^2 - 12x + 32}{(2x-6)^2}$$

5. Critical Numbers and Extrema (continued):

$f'(x) \neq 0$, it is > 0 since Discriminant of numerator
$D = 144 - 4 \cdot 2 \cdot 32 < 0$
$\boxed{f(x) \text{ increases}}$ for all x on the domain

$f'(x)$ DNE $\Rightarrow x = 3$ but not on domain, not a critical number. No change in increasing/decreasing around $x = 3$ because $f(x)$ increases.

$\boxed{\text{There are no critical numbers, no local extrema.}}$

6. Points of Inflection and Concavity:

$$f''(x) = \left(\frac{2x^2 - 12x + 32}{(2x-6)^2}\right)' = \left(\frac{2(x^2 - 6x + 16)}{4(x-3)^2}\right)'$$

$$= \frac{1}{2} \cdot \frac{(x^2-6x+16)'(x-3)^2 - ((x-3)^2)'(x^2-6x+16)}{(x-3)^4}$$

$$= \frac{1}{2} \cdot \frac{(2x-6)(x-3)^2 - 2(x-3)(x^2-6x+16)}{(x-3)^4}$$

$$= \frac{(x-3)^3 - (x-3)(x^2-6x+16)}{(x-3)^4} = -\frac{7}{(x-3)^3}$$

6. Points of Inflection and Concavity (continued):

$f''(x) \neq 0$ or $f''(x)$ DNE
$(x-3)^3 = 0 \Rightarrow x = 3$

$x = 3$ is not a PI, not on the domain, but concavity could change around it.

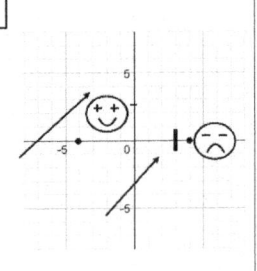

$\boxed{\text{Concave up on } (-\infty, 3) \\ \text{Concave down on } (3, \infty)}$

7. Asymptotes:

Power of numerator is greater than power of denominator, so use long division to find the slant asymptote.

$\boxed{\text{No horizontal asymptote.}}$

$\boxed{\text{The slant asymptote is } y = 0.5x + 1.5}$

Vertical asymptote: Occurs where denominator of $f(x)$ is 0.
$2x - 6 = 0$
$\boxed{\text{Vertical asymptote is } x = 3}$

$$\begin{array}{r} .5x + 1.5 \quad R: -7 \\ 2x-6 \overline{) x^2 - 16} \\ \underline{x^2 - 3x} \\ 3x - 16 \\ \underline{3x - 9} \\ -7 \end{array}$$

$\boxed{f(x) = \dfrac{x^2 - 16}{2x - 6}}$

$$\frac{x^2 - 16}{2x - 6} = 0.5x + 1.5 - \frac{7}{2x - 6}$$

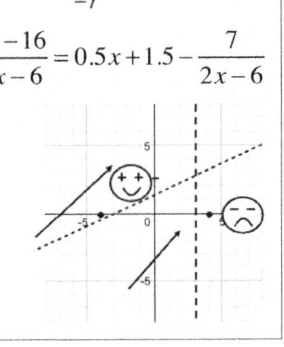

Time to graph!

Check using graphing calculator:

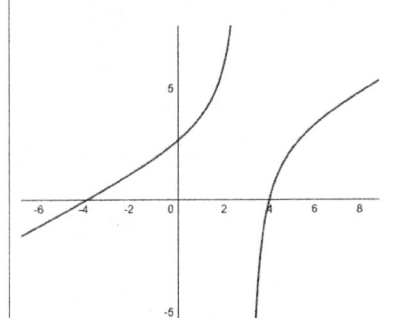

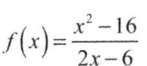

$f(x) = \dfrac{x^2 - 16}{2x - 6}$

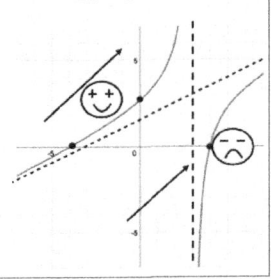

3.6 Optimization Problems

Optimization. Applications of Maxima and Minima

The First Derivative may be used to find the largest or smallest value of a function.

Example 1: Find two positive numbers whose sum is 20 and whose product is as large as possible.

Let the 1st number = x
Then 2nd number = $(20-x)$
product $F(x) = x(20-x) = 20x - x^2$
$F'(x) = 20 - 2x = 2(10-x)$
$F'(x) = 0 \Rightarrow x = 10$

Compare the product $F(x)$ at the endpoints and at $x = 10$:
$F(0) = 0$
$F(10) = 100$ — Maximum
$F(20) = 0$

1st number = 10,
2nd number = 20-10 = 10

Example 2: A rectangle is inscribed in a semicircle of radius 2cm. What is the largest area the rectangle can have and its dimensions?

Let length = $2x \Rightarrow$ Height = $\sqrt{4-x^2}$
Area $A(x) = 2x\sqrt{4-x^2}$

$$A'(x) = \left(2x\sqrt{4-x^2}\right)' = 2\left(x'\sqrt{4-x^2} + \left(\sqrt{4-x^2}\right)' x\right)$$

$$= 2\left(\sqrt{4-x^2} + \frac{1\cdot(-2x)}{2\sqrt{4-x^2}}\cdot x\right)$$

$$= 2\left(\frac{(4-x^2) - x^2}{\sqrt{4-x^2}}\right) = 2\left(\frac{4-2x^2}{\sqrt{4-x^2}}\right) = 0$$

Example 2 Continued

$4 - 2x^2 = 0$
$x = \pm\sqrt{2} \Rightarrow$ Since $0 \le x \le 2$, then $x = \sqrt{2}$

Compare the values for the area at the endpoints and at $x = \sqrt{2}$
$A(0) = 0$,
$A(2) = 0$
$A(\sqrt{2}) = 2\sqrt{2}\cdot\sqrt{2} = 4$

$A(\sqrt{2}) = 4\,cm^2$ is the largest area.
Rectangle is $2\sqrt{2}$ by $\sqrt{2}$

Example 3: A square sheet of tin *16* inches on a side is to make an open-top box by cutting a small square of tin from each corner and bending up the sides. How large a square should be cut from each corner to make the box have as large a volume as possible.

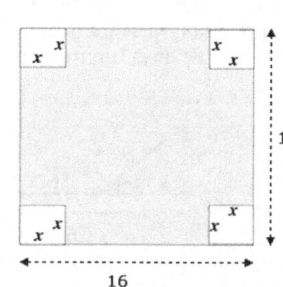

$V(x) = x(16-2x)^2$
$V(x) = x(16^2 - 64x + 4x^2)$
$V(x) = 256x - 64x^2 + 4x^3$
$V'(x) = 256 - 128x + 12x^2$
$= (16-6x)(16-2x)$
$x = 8, \dfrac{8}{3}$

$0 < x < 8$ so $x = \dfrac{8}{3}$

$V(0) = 0, V(8) = 0, V\left(\dfrac{8}{3}\right) > 0$

Dimensions are 8/3 by 8/3

Example 4: A cylindrical metal jar, open at the top, is to have a capacity of 36π in³. The cost of the material used for the bottom of the jar is 10 cents per in², and that of the material used for the curved part is 5 cents per in². What are dimensions that will minimize the cost of the material.

Solution

Cost of container = 10 (area of base) + 5 (lateral area)

$C = 10(\pi r^2) + 5(2\pi rh) = 10\pi(r^2 + rh)$

Express C in one variable: $\pi r^2 h = 36\pi \Rightarrow h = \dfrac{36}{r^2}$

Plug into C: $C = 10\pi\left(r^2 + r \cdot \dfrac{36}{r^2}\right) = 10\pi\left(r^2 + \dfrac{36}{r}\right)$

Find derivative: $C' = 10\pi\left(2r - \dfrac{36}{r^2}\right) = 20\pi\left(r - \dfrac{18}{r^2}\right)$

Set derivative = 0

Example 4 Continued

$C' = 0$

$20\pi\left(r - \dfrac{18}{r^2}\right) = 0$

$r^3 = 18$

$r = \sqrt[3]{18} \approx 2.62$

Check if r = 2.62 is at the absolute minimum

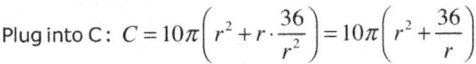

r = 2.62 is at the absolute minimum because the function is decreasing and then increasing

$h = \dfrac{36}{r^2} = \dfrac{36}{2.62^2} \approx 5.24$

Radius by height: 2.62 in by 5.24 in

Example 5: Find the maximum volume of a right circular cylinder that can be inscribed in a right circular cone of altitude 24 centimeters, and base radius 6 centimeters, if the axes of the cylinder and cone coincide.

Volume of cylinder: $V = \pi r^2 h$

Express V in one variable r.

Use similar triangles:

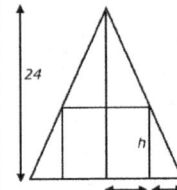

$\dfrac{h}{6-r} = \dfrac{24}{6} = 4 \Rightarrow h = 4(6-r)$

Plug in:

$V = \pi r^2 \cdot 4(6-r) = 4\pi(6r^2 - r^3)$

Example 5 Continued

Differentiate:

$V' = 4\pi(12r - 3r^2)$

$= 4\pi \cdot 3r(4-r) = 12\pi r(4-r)$

Set $V' = 0$

$12\pi r(4-r) = 0$

$r = 4$ and $r \neq 0$

$V(4) = \pi \cdot 4^2 \cdot 4(6-4) = \boxed{128\pi}$

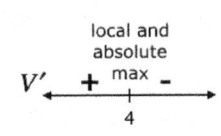

local and absolute max

Example 6

At 9 AM, Fin started biking north from Mathland at 12km/h. At the same time, Inty is is 5km east of Mathland biking west at 16km/h. Express the distance d between Fin and Inty as a function of the time t in hours after 9 AM. At what time are Fin and Inty closest to each other and what is the minimum distance?

$d(t) = ?$ $t_{\min} = ?$ $d_{\min} = ?$

Solution

$d(t) = \sqrt{(12t)^2 + (5-16t)^2}$

$= \sqrt{144t^2 + 25 - 160t + 256t^2}$

$\boxed{d(t) = \sqrt{400t^2 - 160t + 25}}$

Example 6 Continued

Find the minimum of the function using d'(t):

$d'(t) = \dfrac{1}{2\sqrt{400t^2 - 160t + 25}} \cdot (800t - 160)$

$d'(t) = 0$ or $d'(t)$ DNE

$800t - 160 = 0$ Discriminant $D = b^2 - 4ac < 0$

$t = \dfrac{1}{5}$ hours = 12 min $\Rightarrow$ Denominator $\neq 0$

local and absolute min

$d\left(\dfrac{1}{5}\right) = \sqrt{400 \cdot \left(\dfrac{1}{5}\right)^2 - 160 \cdot \left(\dfrac{1}{5}\right) + 25} = \sqrt{16 - 32 + 25} = 3$

Fin and Inty are closest at 9:12 AM   Minimum distance is 3 km

3.7 Rectilinear Motion

Mathboat.com

Rectilinear Motion

If a point P is moving along a line l, its motion is **rectilinear**.

Definitions

Let $s(t)$ be the coordinate of a point P on a coordinate line l at time t.

The velocity of P is $v(t) = s'(t)$

The speed of P is $|v(t)|$.

The acceleration of P is $a(t) = v'(t) = s''(t)$.

$v(t) > 0 \Rightarrow s'(t) > 0 \Rightarrow s(t)$ is increasing
$\Rightarrow$ point P is moving in the positive direction on l
$v(t) < 0 \Rightarrow s'(t) < 0 \Rightarrow s(t)$ is decreasing
$\Rightarrow$ point P is moving in the negative direction on l
$v(t) = 0$ where point P changes direction
$a(t) = v'(t) > 0 \Rightarrow v(t)$ is increasing
$a(t) = v'(t) < 0 \Rightarrow v(t)$ is decreasing

Example 1

The position function s of a point P on a coordinate line is given by $s(t) = t^3 - 12t^2 + 36t + 10$ with t in seconds and $s(t)$ in centimeters. When is point P moving to the right during time interval $[1, 8]$?

Point P is moving to the right $\Rightarrow$ velocity is positive

$v(t) = s'(t) = \left(t^3 - 12t^2 + 36t + 10\right)'$
$= 3t^2 - 24t + 36 = 0$

Example 1 Continued

$3(t - 2)(t - 6) = 0$

$t = 2$ and $t = 6$

$v(t) = s'(t)$

Point P is moving to the right for $t \in [1, 2) \cup (6, 8]$

Example 2

A projectile is fired straight upward. Its distance (in feet) above the ground after t seconds is $s(t) = -16t^2 + 320t$.

a) Find time, velocity and speed at which the projectile hits the ground.

Projectile hits the ground:
$s(t) = -16t^2 + 320t = 0$
$-16t(t - 20) = 0 \Rightarrow t = 0$ and $\boxed{t = 20}$
$v(t) = s'(t) = -32t + 320$
$v(20) = -32(20) + 320 = \boxed{-320 \text{ ft/sec}}$
$v(20) < 0 \Rightarrow$ projectile was moving downwards
The speed at $t = 20$ is $|v(20)| = |-320| = \boxed{320 \text{ ft/sec}}$

Example 2 Continued

$s(t) = -16t^2 + 320t$.

b) Find the maximum altitude achieved by the projectile.
c) Find the acceleration at any time t.

b) To find Max altitude set $s'(t) = 0$
$s'(t) = v(t) = -32t + 320 = 0$
$t = 10$

$s'(t)$  $\underset{10}{\overset{\text{max}}{+\; \cdot \;-}}$

Max altitude is
$s(10) = -16(10)^2 + 320(10) = \boxed{1600\, ft}$

Local max = absolute max since $s(t)$ increases, then decreases

c) The acceleration at any time t is
$a(t) = v'(t) = \boxed{-32\, ft/sec^2}$.

This constant acceleration is caused by the force of gravity.

Velocity

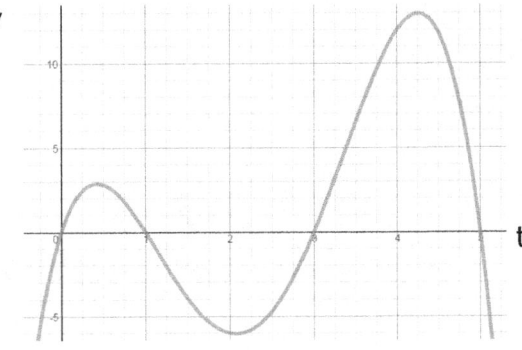

Speed is the absolute value of velocity.

When the velocity graph is moving away from the t-axis, or absolute value of the velocity increases, the speed increases too

---

Now let's graph the Speed. The graph of Speed is the one that is above the x-axis.

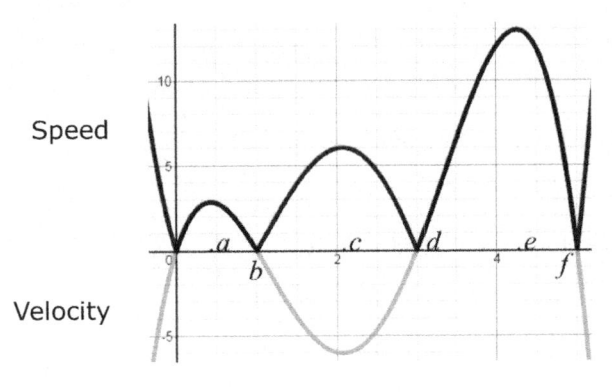

Increasing/Decreasing Speed

Let's fill out this table based on the signs of the Velocity and Acceleration (which is the slope of the Velocity). Then conclude if Speed is increasing or decreasing on each interval.

Interval	Velocity + or -	Acceleration + or -	Speed incr/decreasing
[0,a]	positive	positive	Increasing
[a,b]	positive	negative	Decreasing
[b,c]	negative	negative	Increasing
[c,d]	negative	positive	Decreasing
[d,e]	positive	positive	Increasing
[e,f]	positive	negative	Decreasing

How to determine if the speed is increasing or decreasing:

- If the **velocity and acceleration have the same sign**, then the **speed is increasing**.
- If **velocity and acceleration have different signs**, the **speed is decreasing**.
- If the velocity graph is moving away from the t-axis the speed is increasing.
- If the velocity graph is moving toward the t-axis the speed is decreasing.

Example 3

A particle moves along the x-axis so that its velocity at time t, for $0 \le t \le 6$, is given by a differentiable function v whose graph is shown above. On the interval $2 < t < 3$, is speed of the particle increasing or decreasing?

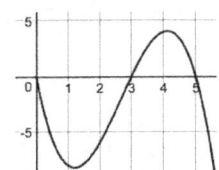

Example 3 Continued

The speed is decreasing on the interval $2 < t < 3$ since on this interval:

1) Velocity $v(t) < 0$ and
2) $v(t)$ is increasing $\Rightarrow v'(t) = a(t) > 0$

So, Velocity and Acceleration have different signs on this interval.

OR: Velocity graph is moving toward the t-axis
$\Rightarrow$ speed decreases

Calculus and Problems that occur in economics.

Cost function: $C(x) =$ cost of producing x units
Average Cost function: $c(x) = \dfrac{C(x)}{x}$
$\quad =$ average cost of producing one unit
Revenue function: $R(x) =$ revenue received for selling x units
Profit function: $P(x) = R(x) - C(x) =$ profit in selling x units

We regard x as a real number, even though this variable may take on only integer values. We always assume $x \geq 0$, since the production of a negative number of units has no practical significance.

Example 4

A manufacturer of sport equipment parts has a monthly fixed cost of $10,000, a production cost of $10 per part, and a selling price of $18 per part.

a) Find $C(x), c(x), R(x)$, and $P(x)$.

The production costs of manufacturing x parts is $10x$.
The total monthly cost $C(x)$ of manufacturing x parts is
$C(x) = \boxed{10x + 10,000}$
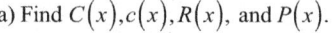
$c(x) = \dfrac{C(x)}{x} = \boxed{10 + \dfrac{10,000}{x}}$
$R(x) = \boxed{18x}$
$P(x) = R(x) - C(x) = 18x - (10x + 10,000) = \boxed{8x - 10,000}$

Example 4 Continued

b) How many parts must be manufactured in order to break even?

The break-even point corresponds to a zero profit:
$P(x) = 0$
$8x - 10000 = 0$
$8x = 10000$, or $x = 1250$

To break even it is necessary to produce and sell 1250 parts per month

Example 5

The weekly cost (in dollars) of manufacturing x wooden tables is given by $C(x) = 1.5x^2 - 960x + 10$. Each table produced is sold for $300. What weekly production rate will maximize the profit?

Profit $= P(x) = R(x) - C(x)$
$= 300x - (1.5x^2 - 960x + 10)$
$= -1.5x^2 + 1260x - 10$

To maximize it, find $P'(x)$ and set it $= 0$

Example 5 Continued

$P'(x) = -3x + 1260 = 0$
$\quad -3(x - 420) = 0$
$\quad x = 420$
$P''(x) = -3 \Rightarrow P''(420) = -3 < 0$
By the Second Derivative test,

a maximum profit occurs if 420 tables per week are manufactured and sold.

Note that $P(420)$ is both local and absolute maximum because $P(x)$ first increases, then decreases.

4.1 Antiderivatives and Indefinite Integrals

Definition

A function $F(x)$ is an antiderivative of $f(x)$ on an interval I if $F'(x) = f(x)$ for every x in I

$F(x) = x^2$ is the antiderivative of $f(x) = 2x$ because
$$F'(x) = (x^2)' = 2x = f(x)$$
However,
$$(x^2)' = (x^2+2)' = (x^2+7)' = (x^2+C)' = 2x$$
So, $(F(x)+C)' = f(x)$ where C is any Constant
Then, $F(x) = ???$

What is antiderivative $F(x)$?

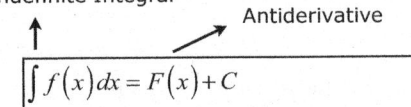

$$\int f(x)\,dx = F(x) + C$$
where $F'(x) = f(x)$ and C is a constant

or: $\int f'(x)\,dx = f(x) + C$

For Example

$\int x^4\,dx = \dfrac{1}{5}x^5 + C$ because $\left(\dfrac{1}{5}x^5\right)' = x^4$

Why is C Important?

C is important because for different C, we get different functions

For Example, $y = \int 3x^2\,dx = x^3 + C$

$C = 0 \Rightarrow y = x^3 + 0$
$C = 1 \Rightarrow y = x^3 + 1$
$C = 2 \Rightarrow y = x^3 + 2$
$C = -1 \Rightarrow y = x^3 - 1$
$C = -2 \Rightarrow y = x^3 - 2$

Integration Formulas

Derivative $f'(x)$ is known	Let's find Indefinite Integral $\int f'(x)\,dx = f(x) + C$
$x' = 1$	$\int 1\,dx = \int dx = x + C$
$\left(\dfrac{x^{r+1}}{r+1}\right)' = x^r \ (r \neq -1)$	$\int x^r\,dx = \dfrac{x^{r+1}}{r+1} + C \ (r \neq -1)$
$(\sin x)' = \cos x$	$\int \cos x\,dx = \sin x + C$
$(-\cos x)' = \sin x$	$\int \sin x\,dx = -\cos x + C$

More Integration Formulas

Derivative $f'(x)$ is known	Let's find Indefinite Integral $\int f'(x)\,dx = f(x) + C$
$(\tan x)' = \sec^2 x$	$\int \sec^2 x\,dx = \tan x + C$
$(-\cot x)' = \csc^2 x$	$\int \csc^2 x\,dx = -\cot x + C$
$(\sec x)' = \sec x \tan x$	$\int \sec x \tan x\,dx = \sec x + C$
$(-\csc x)' = \csc x \cot x$	$\int \csc x \cot x\,dx = -\csc x + C$

Examples

$$\int x^r dx = \frac{x^{r+1}}{r+1} + C \quad (r \neq -1)$$

$$\int \sin x \, dx = -\cos x + C$$

$$\int x^4 \cdot x^7 \, dx = \int x^{11} \, dx = \frac{x^{11+1}}{11+1} + C = \boxed{\frac{x^{12}}{12} + C}$$

$$\int \frac{1}{x^4} dx = \int x^{-4} dx = \frac{x^{-4+1}}{-4+1} + C = \boxed{-\frac{1}{3x^3} + C}$$

$$\int \frac{\tan x}{\sec x} dx = \int \cos x \frac{\sin x}{\cos x} dx = \int \sin x \, dx = \boxed{-\cos x + C}$$

Theorem

I. $\quad \int f'(x) dx = f(x) + C$

(by definition of Indefinite Integral)

II. $\quad \left[\int f(x) dx \right]' = f(x)$

Proof:

$$\left[\int f(x) dx \right]' = \left[F(x) + C \right]' = F'(x) + 0 = f(x)$$

Properties of Indefinite Integral

I. For any Constant c, $\int cf(x) dx = c \int f(x) dx$

II. $\int [f(x) + g(x)] dx = \int f(x) dx + \int g(x) dx$ **Proof**

Let $F(x)$ = antiderivative of $f(x) \Rightarrow F'(x) = f(x)$
and $G(x)$ = antiderivative of $g(x) \Rightarrow G'(x) = g(x)$

$(F(x) + G(x))' = F'(x) + G'(x) = f(x) + g(x)$

So, $\int [f(x) + g(x)] dx = F(x) + G(x) = \int f(x) dx + \int g(x) dx$

III. $\int [f(x) - g(x)] dx = \int f(x) dx - \int g(x) dx$

Example 1

Evaluate

$\int (6x^4 + 3\cos x) dx$ \quad Split Integral into two

$= \int 6x^4 dx + \int 3\cos x \, dx$ \quad Take out Constants

$= 6 \int x^4 dx + 3 \int \cos x \, dx$ \quad Integrate each Term

$= \boxed{\frac{6x^5}{5} + 3\sin x + C}$

Example 2

Evaluate

$\int \frac{(x^2 - 1)^2}{x^2} dx$ \quad Expand $(x^2 - 1)^2$

$= \int \frac{x^4 - 2x^2 + 1}{x^2} dx$ \quad Divide every term of numerator by denominator, x^2

$= \int (x^2 - 2 + x^{-2}) dx$ \quad Integrate each term

$= \frac{x^3}{3} - 2x + \frac{x^{-1}}{-1} + C$

$= \boxed{\frac{1}{3}x^3 - 2x - \frac{1}{x} + C}$

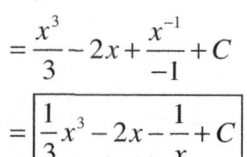

Example 3

Evaluate $\quad \int \frac{5}{\cos u \cot u} du$

$= 5 \cdot \int \frac{1}{\cos u} \cdot \frac{1}{\cot u} du$

$= 5 \int \sec u \tan u \, du$

$= \boxed{5 \sec u + C}$

4.2 Differential Equations

Mathboat.com

Differential Equations

Differential equation is an equation that involves derivatives of an uknown function.

A function f is a solution of a differential equation if it satisfies the equation.
To solve a differential equation means to find all solutions.
When derivative $f'(x)$ is given, we integrate to find the set of functions: $\int f'(x)dx = f(x)+C$
It is called a **general solution** to differential equation.
Sometimes we may know certain values of f or f', called **initial conditions**. It helps us to find the **particular solution**.

Example 1

Find the equation of the curve that passes through $(1,-1)$ and whose slope at any point (x,y) is $3x^2$

Step One - Integrate $y = \int 3x^2 \, dx$

$y = x^3 + C$ general solution

Step Two - Substitute initial conditions $-1 = 1^3 + C$

Step Three - Solve for C $C = -2$

Step Four - Substitute C back into general solution

$\boxed{y = x^3 - 2}$ particular solution

Finding the general solution

Example 2 Solve the differential equation:

$\dfrac{dy}{dx} = x^2\sqrt{y} \Rightarrow \dfrac{dy}{dx} = \dfrac{x^2}{y^{-\frac{1}{2}}}$ Move y to denominator. The exponent of y will become negative

$\int y^{-\frac{1}{2}} dy = \int x^2 dx$ Separate variables and then Integrate

$\dfrac{y^{\frac{1}{2}}}{\frac{1}{2}} = \dfrac{x^3}{3} + C$ add C on one side of the equation, usually the one where x is

$\sqrt{y} = \dfrac{x^3}{6} + C$ Multiply both sides by $\dfrac{1}{2}$
Note: $C \cdot \dfrac{1}{2}$ = Constant, so keep it called C

$y = \left(\dfrac{x^3}{6} + C\right)^2$ Solve for y

Example 3

Solve the differential equation:

$\dfrac{dy}{dx} = \dfrac{x+1}{y-1}; y > 1 \Rightarrow y - 1 > 0$

$\int (y-1)dy = \int (x+1)dx$ Separate variables and Integrate

$\dfrac{y^2}{2} - y = \dfrac{x^2}{2} + x + C$ Integrate **each term of the integrand**

$y^2 - 2y + 1 = x^2 + 2x + 1 + C$ Complete the Square

$\boxed{(y-1)^2 = (x+1)^2 + C}$

Example 3 Continued

OR: Integrate the **whole integrands, not each term**

$\boxed{\dfrac{(y-1)^2}{2} = \dfrac{(x+1)^2}{2} + C}$ Multiply by 2

$(y-1)^2 = (x+1)^2 + C$

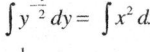

$y - 1 = \pm\sqrt{(x+1)^2 + C}$

Since given that $y > 1$,
then $y - 1$ cannot be negative

So $y - 1 = \sqrt{(x+1)^2 + C}$

Solve for y   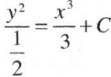 $\boxed{y = 1 + \sqrt{(x+1)^2 + C}}$

Finding the particular solution

Example 4

Find the position s as a function of time t from the given velocity $v(t) = 4t^3$ if $s(0)=4$

Given: $v(t) = 4t^3; s(0) = 4 \quad s(t) = ?$

Solution: $s(t) = \int v(t)\,dt$ — since Velocity is the Derivative of Position

$s(t) = \int 4t^3\,dt$

$s(t) = t^4 + C$ — Add C. This is a general solution

Use initial conditions:

$4 = (0)^4 + C$ — Plug C into general solution

$C = 4$

$\boxed{s(t) = t^4 + 4}$ — This is the particular solution

Example 5

What is the solution to the differential equation

$\dfrac{dy}{dx} = 3x^2 + 2x + 1$

if it given that $y(1) = 0$?

Given:

$\dfrac{dy}{dx} = 3x^2 + 2x + 1$

$y(1) = 0$

$y = ?$

Example 5 Solution

Solution

$\int dy = \int (3x^2 + 2x + 1)\,dx$ — Separate variables and integrate

$y = x^3 + x^2 + x + C$ — This is a general solution

Plug in initial conditions $y(1) = 0$:

$0 = (1)^3 + (1)^2 + 1 + C$ — Plug C into general solution

$C = -3$

$y = x^3 + x^2 + x - 3$ — This is the particular solution

Example 6.

A projectile is fired straight up from a platform 10 ft above the ground, with an initial velocity of 160 ft/sec. Its downward acceleration is 32 ft/sec². Find an equation for the height of the projectile above the ground as a function of time t if $t = 0$ when the projectile is fired.

Example 6 Solution

Given:

$a(t) = -32\,ft/\sec^2$

$v(0) = 160\,ft/\sec$

$s(0) = 10\,ft$

$S(t) = ?$

$v(t) = \int a(t)\,dt = \int -32\,dt$

$= -32t + C$

Use Initial Conditions for v:

$160 = -32 \cdot 0 + C \Rightarrow C = 160$

Plug $C = 160$ into $v(t)$

$\boxed{v(t) = -32t + 160}$

Example 6 Continued

$s(t) = \int v(t)\,dt = \int (-32t + 16)\,dt$

$s(t) = -16t^2 + 160t + C$

Use Initial Conditions for s:

$10 = 0 + 0 + C \Rightarrow C = 10$

Plug $C = 10$ into $s(t)$

$\boxed{s(t) = -16t^2 + 160t + 10}$

4.3 Change of Variables in Indefinite Integrals or Method of Substitution.

Mathboat.com

Method of Substitution

Example 1

Evaluate:

$\int \sqrt{6x+5}\, dx$ — Express in only one variable u

$= \dfrac{1}{6}\int u^{\frac{1}{2}}\, du$ — Substitute. Use the Exponent 1/2 instead of the Square Root

$= \dfrac{1}{6} \dfrac{u^{\frac{3}{2}}}{\frac{3}{2}} + C$ — Integrate

$\boxed{= \dfrac{1}{9}\sqrt{(6x+5)^3} + C}$ — Replace u back with 6x+5

$u = 6x+5$	Set the inner function equal to u
$du = 6dx$	Take the Derivative
$\dfrac{1}{6}du = dx$	Solve for dx

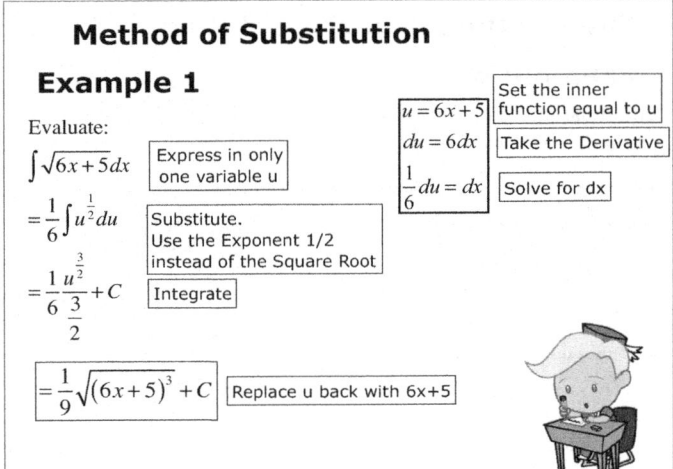

Example 2

Evaluate:

$\int (3x^3+2)^5 x^2\, dx$

$= \dfrac{1}{9}\int u^5\, du$

$= \dfrac{1}{9}\dfrac{u^6}{6} + C$

$\boxed{= \dfrac{1}{54}(3x^3+2)^6 + C}$

Substitution

$u = 3x^3+2$	Set the inner function equal to u
$du = 9x^2\, dx$	Take the Derivative
$\dfrac{du}{9} = x^2\, dx$	Solve for $x^2 dx$

Example 3

Evaluate:

$\int x\sqrt[3]{3-8x^2}\, dx$

$= -\dfrac{1}{16}\int u^{\frac{1}{3}}\, du$

$= -\dfrac{1}{16}\dfrac{u^{\frac{4}{3}}}{\frac{4}{3}} + C$

$= -\dfrac{3}{64}\sqrt[3]{u^4} + C$

$\boxed{= -\dfrac{3}{64}\sqrt[3]{(3-8x^2)^4} + C}$

Substitution

$u = 3-8x^2$	Set the inner function equal to u
$du = -16x\, dx$	Take the Derivative
$-\dfrac{du}{16} = x\, dx$	Solve for $x\, dx$

Example 4

Evaluate:

$\int \dfrac{x^2+1}{(x^3+3x+4)^5}\, dx$

$= \dfrac{1}{3}\int u^{-5}\, du$

$= \dfrac{1}{3}\cdot \dfrac{u^{-4}}{-4} + C$

$= -\dfrac{1}{12}u^{-4} + C$

$\boxed{= -\dfrac{1}{12(x^3+3x+4)^4} + C}$

$u = x^3+3x+4$	Set the inner function of denominator equal to u
$du = (3x^2+3)dx$ $= 3(x^2+1)dx$	Take the Derivative and simplify
$\dfrac{du}{3} = (x^2+1)dx$	Solve for $(x^2+1)dx$

Change of Variables in Integrals of Trigonometric Functions

The Basics

$$\int \cos x \, dx = \sin x + C$$

$$\int \sin x \, dx = -\cos x + C$$

Method of Substitution.

Example 5

Evaluate:

$$\int \cos(2x) \, dx$$
$$= \int \frac{1}{2} \cos u \, du$$
$$= \frac{1}{2} \sin u + C$$
$$= \boxed{\frac{1}{2} \sin(2x) + C}$$

Substitution

$u = 2x$ — Set the inner function equal to u
$du = 2dx$ — Take the Derivative
$\frac{1}{2} du = dx$ — Solve for dx

Let's review Integrals of other Trigonometric Functions

$$\int \sec^2 u \, du = \tan u + C$$

$$\int \csc^2 u \, du = -\cot u + C$$

$$\int \sec u \tan u \, du = \sec u + C$$

$$\int \csc u \cot u \, du = -\csc u + C$$

Example 6

Evaluate:

$$\int \tan^3 x \sec^2 x \, dx$$
$$= \int u^3 \, du$$
$$= \frac{1}{4} u^4 + C$$
$$= \boxed{\frac{1}{4} \tan^4 x + C}$$

Substitution

$u = \tan x$ — Set tan **x** (without the exponent) equal to u
$du = \sec^2 x \, dx$ — Take the Derivative

Example 7

Evaluate:

$$\int \sin^2 x \, dx$$
$$= \int \frac{1 - \cos 2x}{2} \, dx \quad \text{Substitute}$$
$$= \int \frac{1}{2} \, dx - \int \frac{\cos 2x}{2} \, dx \quad \text{Split into two integrals}$$
$$= \frac{x}{2} - \frac{1}{2} \frac{\sin 2x}{2} + C \quad \text{Integrate}$$
$$= \frac{x}{2} - \frac{1}{4} \sin 2x + C$$

Use Trigonometric Formula:

$$\sin^2 x = \frac{1 - \cos 2x}{2}$$

Revisiting Example 1

Using U-substitution:

$$\int \sqrt{6x+5} \, dx$$
$$= \frac{1}{6} \int u^{\frac{1}{2}} \, du$$
$$= \frac{1}{6} \frac{u^{\frac{3}{2}}}{\frac{3}{2}} + C \quad \boxed{= \frac{1}{9} \sqrt{(6x+5)^3} + C}$$

$u = 6x + 5$
$du = 6dx$
$\frac{1}{6} du = dx$

Shortcut for Method of Substitution. Revisiting Example 1

Using the shortcut:
$$\int \sqrt{6x+5}\, \frac{d(6x+5)}{6} =$$
$$\int \sqrt{6x+5}\, \frac{d(6x)}{6} =$$
$$\frac{1}{6}\int \sqrt{6x+5}\, d(6x)$$

Multiply "x" in dx by 6 and divide the whole integral by 6

This way, we can integrate directly without changing the variables

Why adding 5 does not make any difference to the answer?

-Because d(5), or derivative of 5 is 0

$$\int \sqrt{6x+5}\, dx = \frac{1}{6}\frac{(6x+5)^{3/2}}{3/2} + C \quad \boxed{= \frac{1}{9}\sqrt{(6x+5)^3} + C}$$

Revisiting Example 5

Using U-substitution:

$$\int \cos(2x)\, dx$$

$$\boxed{\begin{aligned} u &= 2x \\ du &= 2dx \\ \frac{1}{2}du &= dx \end{aligned}}$$

$$= \int \frac{1}{2}\cos u\, du$$
$$= \frac{1}{2}\sin u + C$$
$$= \boxed{\frac{1}{2}\sin(2x) + C}$$

Shortcut for Method of Substitution. Revisiting Example 5

Using the shortcut:
$$\int \cos(2x)\, d\frac{2x}{2}$$
$$= \frac{1}{2}\int \cos(2x)\, d(2x)$$
$$= \frac{1}{2}\sin(2x) + C$$

Multiply "x" in dx by 2 and divide the whole integral by 2

So, we can integrate directly without changing the variables

When do we use *shortcut* instead of *u-substitution*?

We use the shortcut if the **inner** function **is linear**.

For example, we can use the shorcut to integrate $6\sin(3\theta)$, $(8x+2)^2$, $2\sec^2(4\theta)$, $\sqrt[3]{7x-9}$, etc...

$$\int 6\sin 3\theta\, d\theta =$$
$$\int 6\sin 3\theta\, d\frac{3\theta}{3} =$$
$$2\int \sin 3\theta\, d(3\theta) =$$
$$\boxed{-2\cos 3\theta + C}$$

Practice Using the Shortcut:

$$\int (8x+2)^2\, dx =$$
$$\int (8x+2)^2\, \frac{d(8x)}{8} =$$
$$\int (8x+2)^2\, \frac{d(8x+2)}{8} =$$
$$\frac{1}{8}\frac{(8x+2)^3}{3} + C =$$
$$\boxed{\frac{(8x+2)^3}{24} + C}$$

Practice Using the Shortcut:

$$\int 2\sec^2 4\theta\, d\theta =$$
$$\int 2\sec^2 4\theta\, d\frac{4\theta}{4} =$$
$$\frac{1}{2}\int \sec^2 4\theta\, d(4\theta) =$$
$$\boxed{\frac{1}{2}\tan 4\theta + C}$$

4.4 Summation Notation and Area

 Mathboat.com

Sigma ($\sum$) notation

The expression $\sum_{k=1}^{n} a_k = a_1 + a_2 + a_3 + ... + a_n$
represents the sum of the numbers $a_1, a_2, a_3, ...$ and a_n.

Example 1

Evaluate: $\sum_{k=1}^{3} k^3(k-3)(k+1)$

$= 1^3(1-3)(1+1) + 2^3(2-3)(2+1) + 3^3(3-3)(3+1)$

$= 1(-2)(2) + 8(-1)(3) + 27(0)(4) = -28$

Formulas
(could be proved with Math Induction)

$\sum_{k=1}^{n} k = 1 + 2 + 3 + ... + n = \dfrac{n(n+1)}{2}$

$\sum_{k=1}^{n} k^2 = 1^2 + 2^2 + 3^2 + ... + n^2 = \dfrac{n(n+1)(2n+1)}{6}$

Proof of this Formula is at end of the lesson:

$\sum_{k=1}^{n} k^3 = 1^3 + 2^3 + 3^3 + ... + n^3 = \left(\dfrac{n(n+1)}{2}\right)^2$

$\sum_{k=1}^{n} (a_k \pm b_k) = \sum_{k=1}^{n} a_k \pm \sum_{k=1}^{n} b_k$

$\sum_{k=1}^{n} ca_k = c\sum_{k=1}^{n} a_k$ for any Real Number c

Example 2
Evaluate

a) $\sum_{k=1}^{50} k = 1 + 2 + ... + 50$

$= \dfrac{50(51)}{2} = \boxed{1275}$

Formula: $\boxed{\sum_{k=1}^{n} k = \dfrac{n(n+1)}{2}}$

b) $\sum_{k=1}^{35} k^2 = 1^2 + 2^2 + ... + 35^2$

$= \dfrac{35(36)(71)}{6} = \boxed{14910}$

Formula: $\boxed{\sum_{k=1}^{n} k^2 = \dfrac{n(n+1)(2n+1)}{6}}$

Example 3
Express in terms of n:

$\sum_{k=1}^{n} (k^2 - 4k + 5)$

$= \sum_{k=1}^{n} k^2 - \sum_{k=1}^{n} 4k + \sum_{k=1}^{n} 5$

$= \dfrac{n(n+1)(2n+1)}{6} - 4\dfrac{n(n+1)}{2} + 5n$

$= \boxed{\dfrac{1}{3}n^3 - \dfrac{3}{2}n^2 + \dfrac{19}{6}n}$

Area Under a Curve

The area under the curve can be approximated by adding the areas of inscribed or circumscribed rectangles.

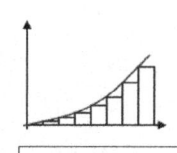

The thinner the rectangles and the more there are, the more exact the area will be.

Inscribed rectangles: Build each rectangle using the lowest height

Circumscribed rectangles: Build each rectangle using the largest height

Example 4

Approximate area under $y = x+1$, from 0 to 4, using both methods. $\Delta x = 1$

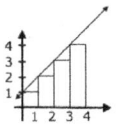 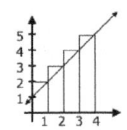

$h_1 = 1$ $\sum \text{areas} =$ $h_1 = 2$ $\sum \text{areas} =$
$h_2 = 2$ $1 \cdot 1 + 1 \cdot 2 +$ $h_2 = 3$ $1 \cdot 2 + 1 \cdot 3 +$
$h_3 = 3$ $1 \cdot 3 + 1 \cdot 4 = 10$ $h_3 = 4$ $1 \cdot 4 + 1 \cdot 5 = 14$
$h_4 = 4$ $h_4 = 5$

Example 5

Approximate area under the curve of $f(x) = 4 - x^2$ on $[0,2]$ by dividing $[0,2]$ into subintervals of equal length $\Delta x = 0.5$ and using a) inscribed rectangles b) circumscribed rectangles

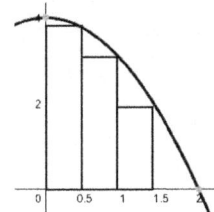 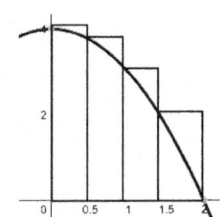

$A_{IP} = 0.5(f(0.5) + f(1) + f(1.5))$ $A_{CP} = 0.5(f(0) + f(0.5) + f(1) + f(1.5))$
$A_{IP} = 0.5(3.75 + 3 + 1.75) = \boxed{4.25}$ $A_{CP} = 0.5(4 + 3.75 + 3 + 1.75) = \boxed{6.25}$

Example 6

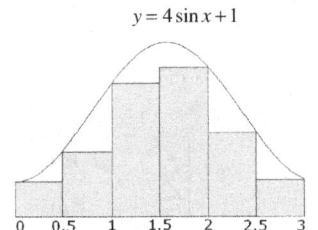

Find the area under the curve $y = 4\sin x + 1$ from $x = 0$ to $x = 3$ using 6 inscribed rectangles

$\Delta x = \dfrac{b-a}{n} = \dfrac{3-0}{6} = 0.5$

Area of rectangle = length · width
Area = $\Delta x (f(0) + f(0.5) + f(1) + f(2) + f(2.5) + f(3))$
Area = $0.5(1 + 2.9 + 4.4 + 4.6 + 3.4 + 1.6)$

$f(0) = 4\sin 0 + 1 = 1$
$f(0.5) = 4\sin 0.5 + 1 \approx 2.9$
$f(1) = 4\sin 1 + 1 \approx 4.4$
$f(2) = 4\sin 2 + 1 \approx 4.6$
$f(2.5) = 4\sin 2.5 + 1 \approx 3.4$
$f(3) = 4\sin 3 + 1 \approx 1.6$

$\boxed{\text{Area} \approx 8.94}$

Example 7

Approximate the area under the curve $y = x^2 - 4x + 4$ using 5 inscribed rectangles of equal width on the interval $[-1, 4]$.

Total Area $= A_1 + A_2 + A_3 + A_4 + A_5$
$A_1 = 1 \cdot ((0)^2 - 4 \cdot (0) + 4) = 4$
$A_2 = 1 \cdot (1^2 - 4 \cdot 1 + 4) = 1$
$A_3 = 0$, No rectangle
$A_4 = 0$, No rectangle
$A_5 = 1 \cdot (3^2 - 4 \cdot 3 + 4) = 1$
Total Area =
$4 + 1 + 0 + 0 + 1 = \boxed{6}$

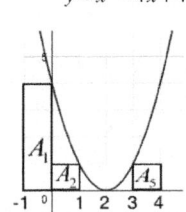 

Prove with Math Induction:

$\sum_{k=1}^{n} k^3 = \left[\dfrac{n(n+1)}{2}\right]^2$ or $1^3 + 2^3 + 3^3 + 4^3 + \ldots + n^3 = \left[\dfrac{n(n+1)}{2}\right]^2$

Step 1: Check if it works for $n = 1$

$1^3 = \left[\dfrac{1(1+1)}{2}\right]^2 = 1$

Step 2: Assume for $n = k$ it is true

$1^3 + 2^3 + 3^3 + 4^3 + \ldots + k^3 = \left[\dfrac{k(k+1)}{2}\right]^2$

Step 3: Prove for $n = k+1$ it is true

$1^3 + 2^3 + 3^3 + 4^3 + \ldots + k^3 + (k+1)^3 = \left[\dfrac{(k+1)(k+2)}{2}\right]^2$

Proof with Math Induction Continued.

Proof: $1^3 + 2^3 + 3^3 + 4^3 + \ldots k^3 + (k+1)^3 =$

$\left[\dfrac{k(k+1)}{2}\right]^2 + (k+1)^3 = \dfrac{k^2(k+1)^2}{4} + (k+1)^3$

$= (k+1)^2 \left[\dfrac{k^2}{4} + \dfrac{k+1}{1}\right] (k+1)^2 \left[\dfrac{k^2 + 4k + 4}{4}\right]$

$= \dfrac{(k+1)^2 (k+2)^2}{2^2}$

Step 4: True for all n

4.5 a
The Definite Integral

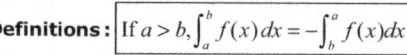
Mathboat.com

Riemann Sum

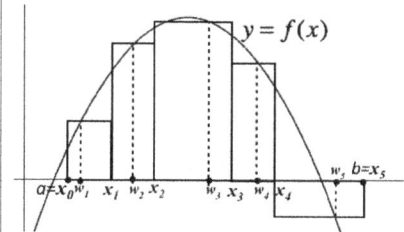

Let's pick some w_k in each subinterval.
Then build the rectangle with height $= f(w_k)$
If $f(w_k) > 0$, rectangle lies above x-axis,
If $f(w_k) < 0$, rectangle lies below x-axis

Riemann Sum continued.

R_p is the sum of areas of rectangles
that lies above x-axis
and negatives of areas of rectangles
that lies below x-axis

Let $f(x)$ be defined on a closed interval $[a,b]$ and P be a partition of $[a,b]$, or P is any decomposition of $[a,b]$ into subintervals of the form $[x_{k-1}, x_k]$.

Riemann sum of $f(x)$ for P is
$R_P = \sum_{k=1}^{n} f(w_k)\Delta x_k$ where $w_k \in [x_{k-1}, x_k]$

The Definite Integral

Definition. The Definite Integral of f from a to b is
$$\int_a^b f(x)\,dx = \lim_{\|p\| \to 0} \sum_k f(w_k)\Delta x_k$$
provided the limit exists, where $\|p\|$ is the norm of the partition (largest Δx_k)

Theorem. If f is integrable and $f(x) \geq 0$ for every x in $[a,b]$ (where $a < b$), then the area A of the region under the graph of f from a to b is $A = \int_a^b f(x)\,dx$

Proof: $f(x) \geq 0, a < b \Rightarrow$

$A = \lim_{\|p\| \to 0} \sum_k f(w_k)\Delta x_k \Rightarrow A = \int_a^b f(x)\,dx$

where dx may be assosiated with
the increment of Riemann sum of f

$A = \int_a^b f(x)\,dx$

Definitions: If $a > b, \int_a^b f(x)\,dx = -\int_b^a f(x)\,dx$

If $f(a)$ exists, then $\int_a^a f(x)\,dx = 0$

Areas of the regions are always positive, but definite integrals could be negative or positive

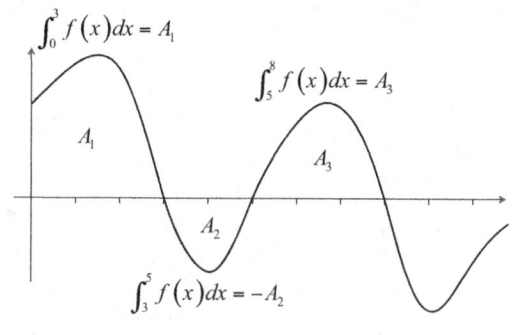

$\int_0^3 f(x)\,dx = A_1$

$\int_5^8 f(x)\,dx = A_3$

$\int_3^5 f(x)\,dx = -A_2$

$\int_0^8 f(x)\,dx = A_1 - A_2 + A_3$

Example 1

Let $f(x) = 8 - 0.5x^2$ and P be partition of $[0,6]$, determined by: $x_0 = 0$, $x_1 = 1.5, x_2 = 2.5, x_3 = 4.5, x_4 = 5$, and $x_5 = 6$. Find a) norm of the partition.

a) $\Delta x_1 = 1.5 - 0 = 1.5, \Delta x_2 = 1, \Delta x_3 = 2, \Delta x_4 = 0.5, \Delta x_5 = 1$.

The norm $\|p\|$ of the partition (largest Δx_k) is $\boxed{\Delta x_3 = 2}$

Example 1 Continued

Let $f(x) = 8 - 0.5x^2$ and P be partition of $[0,6]$, determined by: $x_0 = 0$, $x_1 = 1.5, x_2 = 2.5, x_3 = 4.5, x_4 = 5$, and $x_5 = 6$. Find b) Riemann sum of f by choosing numbers: $w_1 = 1, w_2 = 2, w_3 = 3.5, w_4 = 5, w_5 = 5.5$ in the subintervals of P.

b) $R_p = \sum_{k=1}^{5} f(w_k)\Delta x_k = f(1)(1.5) + f(2)(1) + f(3.5)(2) + f(5)(0.5) + f(5.5)(1)$

where $f(1) = 7.5, f(2) = 6, f(3.5) = 1.875, f(5) = -4.5, f(5.5) = -7.125$

So, $R_p = (7.5)(1.5) + (6)(1) + (1.875)(2) + (-4.5)(0.5) + (-7.125)(1) = \boxed{11.625}$

Example 1 (Continued)

Let $f(x) = 8 - 0.5x^2$ and P be partition of $[0,6]$, determined by: $x_0 = 0$, $x_1 = 1.5, x_2 = 2.5, x_3 = 4.5, x_4 = 5$, and $x_5 = 6$. Find c) Left Riemann sum d) Right Riemann sum e) Midpoint Riemann sum

c) $LRAM = f(0)(1.5) + f(1.5)(1) + f(2.5)(2) + f(4.5)(0.5) + f(5)(1)$

$= \boxed{23.0625}$

where $f(0) = 8, f(1.5) = 6.875, f(2.5) = 4.875,$

$f(4.5) = -2.125, f(5) = -4.5$

Example 1 (Continued)

d) $RRAM = f(1.5)(1.5) + f(2.5)(1) + f(4.5)(2) + f(5)(0.5) + f(6)(1)$

$= 6.875 \cdot (1.5) + 4.875 \cdot (1) + -2.125 \cdot (2) + (-4.5) \cdot (0.5) + (-10) \cdot (1) = \boxed{-1.3125}$

e) $MRAM = f(.75)(1.5) + f(2)(1) + f(3.5)(2) + f(4.75)(0.5) + f(5.5)(1) = \boxed{0.87031}$

where $f(.75) = 7.718, f(2) = 6, f(3.5) = 1.875, f(4.75) = -3.281, f(5.5) = -7.125$

Example 2

Evaluate: $\int_{-5}^{5} \left(\frac{2}{5}x + 4\right)dx$

$\int_{-5}^{5} \left(\frac{2}{5}x + 4\right)dx = A_{trapezoid}$

$= \frac{1}{2}(base_1 + base_2) \cdot height$

$= \frac{1}{2}(2 + 6) \cdot 10 = \boxed{40}$

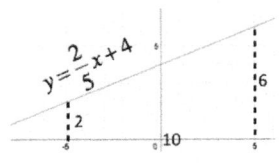

$base_1 = f(-5) = \frac{2}{5} \cdot (-5) + 4 = 2$

$base_2 = f(5) = \frac{2}{5} \cdot (5) + 4 = 6$

Example 3

Evaluate: $\int_{-5}^{5} \sqrt{25 - x^2}\, dx$

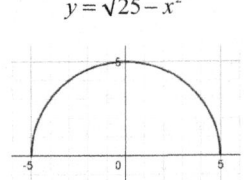

$y = \sqrt{25 - x^2}$

$\int_{-5}^{5} \sqrt{25 - x^2}\, dx = A_{semicircle}$

$= \frac{1}{2} \cdot \pi(5)^2 = \boxed{\dfrac{25\pi}{2}}$

4.5b Limit of Riemann Sum-to-Integral and Integral-to-Limit

Mathboat.com

Example 1 — Limit-to-Integral

If n is a positive integer, then $\lim_{n\to\infty} \dfrac{1}{n}\left[\dfrac{1}{1+\frac{1}{n}}+\dfrac{1}{1+\frac{2}{n}}+\ldots+\dfrac{1}{1+\frac{n}{n}}\right]$ can be:

$\Delta x = \frac{1}{n}$, arrows at x_1, x_2, x_n (from 0 to 1)

(A) $\int_1^2 \dfrac{1}{x}\,dx$ (B) $\int_1^2 \dfrac{1}{x+1}\,dx$
(C) $\int_1^2 x\,dx$ (D) $\int_1^2 \dfrac{2}{x+1}\,dx$ (E) $\int_0^1 \dfrac{1}{x}\,dx$

It looks like a limit of Riemann sums $\lim_{\|p\|\to 0}\sum_{k=1}^{n} f(w_k)\Delta x_k$

From this, we can say $a=0, b=1,$ and $f(x)=\dfrac{1}{1+x}$

That is, the answer is $\int_0^1 \dfrac{1}{1+x}\,dx$. But... no such answer!

Example 1 Continued

Let's try other approach

$\lim_{n\to\infty} \dfrac{1}{n}\left[\dfrac{1}{1+\frac{1}{n}}+\dfrac{1}{1+\frac{2}{n}}+\ldots+\dfrac{1}{1+\frac{n}{n}}\right]$

arrows: $x_1 \to 1$, $x_n \to 2$

So the answer can be $\int_1^2 \dfrac{1}{x}\,dx$

(A) $\boxed{\int_1^2 \dfrac{1}{x}\,dx}$ (B) $\int_1^2 \dfrac{1}{x+1}\,dx$ (C) $\int_1^2 x\,dx$ (D) $\int_1^2 \dfrac{2}{x+1}\,dx$ (E) $\int_0^1 \dfrac{1}{x}\,dx$

Example 2

If n is a positive integer, then $\lim_{n\to\infty}\dfrac{1}{n}\left[\dfrac{5}{2+\frac{1}{n}}+\dfrac{5}{2+\frac{2}{n}}+\ldots+\dfrac{5}{2+\frac{n}{n}}\right]$ can be

$\Delta x = \frac{1}{n}$, arrows at x_1, x_2, x_n (from 0 to 1)

(A) $\int_0^1 \dfrac{5}{x}\,dx$ (B) $\int_1^2 \dfrac{5}{x+2}\,dx$
(C) $\int_0^1 5x\,dx$ (D) $\int_2^3 \dfrac{5}{x}\,dx$

It looks like a limit of Riemann sums $\lim_{\|p\|\to 0}\sum_{k=1}^{n} f(w_k)\Delta x_k$

From this, we can say $a=0, b=1,$ and $f(x)=\dfrac{5}{2+x}$

That is, the answer is $\int_0^1 \dfrac{5}{2+x}\,dx$. But... no such answer!

Example 2 Continued

Let's try other approach

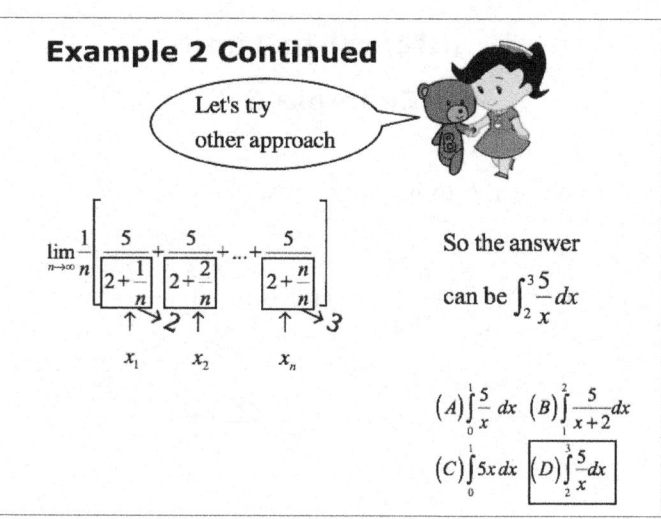

$\lim_{n\to\infty} \dfrac{1}{n}\left[\dfrac{5}{2+\frac{1}{n}}+\dfrac{5}{2+\frac{2}{n}}+\ldots+\dfrac{5}{2+\frac{n}{n}}\right]$

arrows: $x_1 \to 2$, $x_n \to 3$

So the answer can be $\int_2^3 \dfrac{5}{x}\,dx$

(A) $\int_0^1 \dfrac{5}{x}\,dx$ (B) $\int_1^2 \dfrac{5}{x+2}\,dx$
(C) $\int_0^1 5x\,dx$ (D) $\boxed{\int_2^3 \dfrac{5}{x}\,dx}$

Example 3

If n is a positive integer, express $\lim_{n\to\infty}\dfrac{1}{n^4}\left[1^3+2^3\ldots+n^3\right]$ as definite integral.

$\lim_{n\to\infty}\dfrac{1}{n^4}\left[1^3+2^3\ldots+n^3\right] = \lim_{n\to\infty}\dfrac{1}{n}\left[\dfrac{1^3}{n^3}+\dfrac{2^3}{n^3}+\ldots+\dfrac{n^3}{n^3}\right] =$

It looks like a limit of Riemann sums $\lim_{\|p\|\to 0}\sum_{k=1}^{n} f(w_k)\Delta x_k$

$\Delta x = \frac{1}{n}$

$\lim_{n\to\infty}\dfrac{1}{n}\left[\left(\dfrac{1}{n}\right)^3+\left(\dfrac{2}{n}\right)^3+\ldots+\left(\dfrac{n}{n}\right)^3\right] = \boxed{\int_0^1 x^3\,dx}$

arrows at $x_1 \to 0, x_2, x_n \to 1$

Example 4

If n is a positive integer, then $\lim\limits_{n\to\infty}\dfrac{1}{n}\left[\sin\dfrac{\pi}{n}+\sin\dfrac{2\pi}{n}+\ldots+\sin\dfrac{n\pi}{n}\right]$ can be

a) $\int_0^1 \sin\dfrac{\pi}{x}\,dx$ b) $\int_0^1 \cos(\pi x)\,dx$ c) $\int_0^1 \sin(\pi x)\,dx$ d) $\int_0^\pi \sin(\pi x)\,dx$

Example 4 Solution

It looks like a limit of Riemann sums

$$\lim_{\|p\|\to 0}\sum_{k=1}^{n} f(w_k)\Delta x_k$$

$\Delta x = \dfrac{1}{n}$

$$\lim_{n\to\infty}\dfrac{1}{n}\left[\sin\dfrac{\pi\cdot 1}{n}+\sin\dfrac{\pi\cdot 2}{n}+\ldots+\sin\dfrac{\pi\cdot n}{n}\right] = \int_0^1 \sin(\pi x)\,dx$$

$\quad\quad\quad\quad\quad\downarrow 0 \quad\quad\uparrow \quad\quad\quad\quad\downarrow 1$
$\quad\quad\quad\quad\quad x_1 \quad\quad x_2 \quad\quad\quad x_n$

[C]

Limit of Riemann Sum-to-Integral

Example 5 Express as Integral: $\lim\limits_{n\to\infty}\sum\limits_{k=1}^{n}\left(2+\dfrac{7k}{n}\right)^2\left(\dfrac{7}{n}\right) = \lim\limits_{n\to\infty}\sum\limits_{k=1}^{n} f(x_k)\Delta x$

1) Identify x_k and $f(x_k)$	$x_k = 2+\dfrac{7k}{n}$	$\Rightarrow f(x_k) = x_k^2$
2) Find a: let $k=0$ Find b: let $k=n$	$a = x_0 = 2+\dfrac{7\cdot 0}{n} = 2$	$b = x_n = 2+\dfrac{7\cdot n}{n} = 9$
3) Find Δx: $\Delta x = \dfrac{b-a}{n}$	$\Delta x = \dfrac{b-a}{n} = \dfrac{\square-\square}{\square} = \dfrac{\square}{n}$	
4) Write Integral: $\int_a^b f(x)\,dx$	$\int_2^9 x^2\,dx$	

Limit of Riemann Sum-to-Integral

Example 5 continued Express as Integral: $\lim\limits_{n\to\infty}\sum\limits_{k=1}^{n}\left(2+\dfrac{7k}{n}\right)^2\left(\dfrac{7}{n}\right) = \lim\limits_{n\to\infty}\sum\limits_{k=1}^{n} f(x_k)\Delta x$

OR:

1) Identify x_k and $f(x_k)$	$x_k = \dfrac{7k}{n}$	$\Rightarrow f(x_k) = (2+x_k)^2$
2) Find a: let $k=0$ Find b: let $k=n$	$a = x_0 = \dfrac{7\cdot 0}{n} = 0$	$b = x_n = \dfrac{7\cdot n}{n} = 7$
3) Find Δx: $\Delta x = \dfrac{b-a}{n}$	$\Delta x = \dfrac{b-a}{n} = \dfrac{7-0}{n} = \dfrac{7}{n}$	
4) Write Integral: $\int_a^b f(x)\,dx$	$\int_0^7 (2+x)^2\,dx$	

Limit of Riemann Sum-to-Integral

Example 5 continued Express as Integral: $\lim\limits_{n\to\infty}\sum\limits_{k=1}^{n}\left(2+\dfrac{7k}{n}\right)^2\left(\dfrac{7}{n}\right) = \lim\limits_{n\to\infty}\sum\limits_{k=1}^{n} f(x_k)\Delta x$

OR:

1) Identify x_k and $f(x_k)$	$x_k = \dfrac{k}{n}$	$\Rightarrow f(x_k) = (2+7x_k)^2$
2) Find a: let $k=0$ Find b: let $k=n$	$a = x_0 = \dfrac{0}{n} = 0$	$b = x_n = \dfrac{n}{n} = 1$
3) Find Δx: $\Delta x = \dfrac{b-a}{n}$		
4) Write Integral: $\int_a^b f(x)\,dx$	$7\int_0^1 (2+7x)^2\,dx$	

Integral-to-Limit

Example 6

Which of the following limits is equal to $\int_3^5 x^4\,dx$?

(A) $\lim\limits_{n\to\infty}\sum\limits_{k=1}^{n}\left(3+\dfrac{k}{n}\right)^4 \dfrac{1}{n}$ (B) $\lim\limits_{n\to\infty}\sum\limits_{k=1}^{n}\left(3+\dfrac{k}{n}\right)^4 \dfrac{2}{n}$

(C) $\lim\limits_{n\to\infty}\sum\limits_{k=1}^{n}\left(3+\dfrac{2k}{n}\right)^4 \dfrac{1}{n}$ (D) $\lim\limits_{n\to\infty}\sum\limits_{k=1}^{n}\left(3+\dfrac{2k}{n}\right)^4 \dfrac{2}{n}$

Example 6 Solution

By definition of a definite integral,

$$\int_a^b f(x)\,dx = \lim_{n\to\infty} \sum_{k=1}^n f(x_k)\Delta x, \text{ where } [a,b]=[3,5]$$

is divided into n equal subintervals of width

$$\Delta x = \frac{b-a}{n} = \frac{5-3}{n} = \frac{2}{n} \text{ and each}$$

$$x_k = a + k\Delta x = 3 + \frac{2k}{n}$$

(the right endpoint of each subinterval).

Example 6 Continued

Thus, $\int_3^5 x^4\,dx = \lim_{n\to\infty} \sum_{k=1}^n x_k^4 \frac{2}{n} = \lim_{n\to\infty} \sum_{k=1}^n \left(3+\frac{2k}{n}\right)^4 \frac{2}{n}$.

So the short solution is:

$a=3,\ b=5$

$\Delta x = \frac{b-a}{n} = \frac{2}{n}$

$x_k = a + k\Delta x = 3 + \frac{2k}{n}$

$A = \int_3^5 x^4\,dx = \lim_{n\to\infty} \sum_{k=1}^n x_k^4 \frac{2}{n} = \boxed{\lim_{n\to\infty} \sum_{k=1}^n \left(3+\frac{2k}{n}\right)^4 \frac{2}{n}}$ **D**

Example 7

Which of the following limits is equal to $\int_0^4 9x\,dx$?

(A) $\lim_{n\to\infty} \sum_{k=1}^n \left(\frac{k}{n}\right)\frac{9}{n}$ (B) $\lim_{n\to\infty} \sum_{k=1}^n \left(9\left(\frac{4k}{n}\right)\right)\frac{4}{n}$

(C) $\lim_{n\to\infty} \sum_{k=1}^n \left(9\left(\frac{4+k}{n}\right)\right)\frac{1}{n}$ (D) $\lim_{n\to\infty} \sum_{k=1}^n \left(9\left(4+\frac{k}{n}\right)\right)\frac{4}{n}$

By definition of a definite integral, $\int_a^b f(x)\,dx = \lim_{n\to\infty} \sum_{k=1}^n f(x_k)\Delta x$, where $[a,b]=[0,4]$

is divided into n equal subintervals of width $\Delta x = \frac{b-a}{n} = \frac{4-0}{n} = \frac{4}{n}$ and each

$x_k = a + k\Delta x = 0 + \frac{4k}{n}$ (the right endpoint of each subinterval).

Example 7 Continued

Thus, $\int_0^4 9x\,dx = \lim_{n\to\infty} \sum_{k=1}^n 9x_k \frac{4-0}{n} = \lim_{n\to\infty} \sum_{k=1}^n \left(9\left(\frac{4k}{n}\right)\right)\frac{4}{n}$.

So the short solution is:

$a=0,\ b=4$

$\Delta x = \frac{b-a}{n} = \frac{4}{n}$

$x_k = a + k\Delta x = \frac{4k}{n}$

$A = \int_0^4 9x\,dx = \lim_{n\to\infty} \sum_{k=1}^n 9x_k \frac{4}{n} = \boxed{\lim_{n\to\infty} \sum_{k=1}^n \left(9\left(\frac{4k}{n}\right)\right)\frac{4}{n}}$ **B**

Example 8

Compute $\int_1^3 (12-x^2)\,dx$ by writing the integral

as the limit of a Riemann sums.

Solution

By the definition of a definite integral, $\int_a^b f(x)\,dx = \lim_{n\to\infty} \sum_{k=1}^n f(x_k)\Delta x$,

where $[a,b]=[1,3]$ is divided into n equal subintervals of width

$\Delta x = \frac{b-a}{n} = \frac{3-1}{n} = \frac{2}{n}$ and each $x_k = a + k\Delta x = 1 + \frac{2k}{n}$ (the right

endpoint of each subinterval).

Thus, $\int_1^3 (12-x^2)\,dx = \lim_{n\to\infty} \sum_{k=1}^n (12-x_k^2)\frac{2}{n} = \lim_{n\to\infty} \sum_{k=1}^n \left(12-\left(1+\frac{2k}{n}\right)^2\right)\frac{2}{n}$

Example 8 Continued

$= \lim_{n\to\infty} \frac{2}{n} \sum_{k=1}^n \left(12 - 1 - \frac{4k}{n} - \frac{4k^2}{n^2}\right) = 2\lim_{n\to\infty} \sum_{k=1}^n \left(\frac{11}{n} - \frac{4k}{n^2} - \frac{4k^2}{n^3}\right)$

Using formulas: $\sum_{k=1}^n k = \frac{n(n+1)}{2}$ and $\sum_{k=1}^n k^2 = \frac{n(n+1)(2n+1)}{6}$,

$= 2\lim_{n\to\infty} \left(\frac{11n}{n} - \frac{n(n+1)}{2}\cdot\frac{4}{n^2} - \frac{n(n+1)(2n+1)}{6}\cdot\frac{4}{n^3}\right)$

$= 2\lim_{n\to\infty} \left(11 - \frac{\left(1+\frac{1}{n}\right)}{2}\cdot 4 - \frac{\left(1+\frac{1}{n}\right)\left(2+\frac{1}{n}\right)}{6}\cdot 4\right) = 2\left(11 - 2 - \frac{4}{3}\right) = \boxed{\frac{46}{3}}$

4.6 Properties of the Definite Integrals

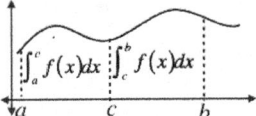

Property: Definite Integral of a Constant Function

If c is real number,
$$\int_a^b c\,dx = c(b-a)$$

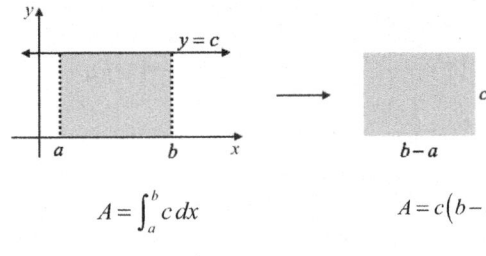

$A = \int_a^b c\,dx \qquad A = c(b-a)$

$$\int_a^b c\,dx = c(b-a)$$

Theorem

If f is integrable on $[a,b]$ and c is any real number, then cf is integrable on $[a,b]$ and $\int_a^b cf(x)\,dx = c\int_a^b f(x)\,dx$

Proof

By definition, $\int_a^b f(x)\,dx = \lim_{\|p\|\to 0} \sum_k f(w_k)\Delta x_k$.

So, $\int_a^b cf(x)\,dx = \lim_{\|p\|\to 0} \sum_k cf(w_k)\Delta x_k = \lim_{\|p\|\to 0} c\sum_k f(w_k)\Delta x_k$

$= c\lim_{\|p\|\to 0} \sum_k f(w_k)\Delta x_k = c\int_a^b f(x)\,dx$

Theorem

If f and g are integrable on $[a,b]$, then $f+g$ and $f-g$ are integrable on $[a,b]$ and:

1. $\int_a^b [f(x)+g(x)]\,dx = \int_a^b f(x)\,dx + \int_a^b g(x)\,dx$
2. $\int_a^b [f(x)-g(x)]\,dx = \int_a^b f(x)\,dx - \int_a^b g(x)\,dx$

Example 1

Simplify:

$\int_0^2 (5x^3 - 3x + 6)\,dx$

$= \int_0^2 5x^3\,dx - \int_0^2 3x\,dx + \int_0^2 6\,dx = 5\int_0^2 x^3\,dx - 3\int_0^2 x\,dx + 6\int_0^2 dx$

Property

If $a < c < b$ and if f is integrable and $f(x) \geq 0$ on both $[a,c]$ and $[c,b]$, then f is integrable on $[a,b]$ and $\int_a^b f(x)\,dx = \int_a^c f(x)\,dx + \int_c^b f(x)\,dx$

Proof

$\int_a^c f(x)dx \quad \int_c^b f(x)dx$

Total area = sum of areas

$$\int_a^b f(x)\,dx = \int_a^c f(x)\,dx + \int_c^b f(x)\,dx$$

This could also be proved for $f(x) < 0$

Property

If f is integrable on a closed interval and if a,b and c are any three numbers in the interval, then $\int_a^b f(x)\,dx = \int_a^c f(x)\,dx + \int_c^b f(x)\,dx$

Proof. Suppose $c < a < b$

$\int_c^b f(x)\,dx = \int_c^a f(x)\,dx + \int_a^b f(x)\,dx$

By previous property, since $c < a < b$

$\int_a^b f(x)\,dx = -\int_c^a f(x)\,dx + \int_c^b f(x)\,dx$

Solve for $\int_a^b f(x)\,dx$. Since $\int_a^c f(x)\,dx = -\int_c^a f(x)\,dx$,

$$\int_a^b f(x)\,dx = \int_a^c f(x)\,dx + \int_c^b f(x)\,dx$$

Example 2

Express as one Integral

$$\int_1^6 f(x)dx - \int_5^6 f(x)dx =$$

Since $\int_a^c f(x)dx = -\int_c^a f(x)dx$,

$$\int_1^6 f(x)dx - -\int_6^5 f(x)dx =$$

Since $\int_a^b f(x)dx + \int_b^c f(x)dx = \int_a^c f(x)dx$,

$$\int_1^6 f(x)dx + \int_6^5 f(x)dx = \boxed{\int_1^5 f(x)dx}$$

Theorem

If f is integrable on $[a,b]$ and $f(x) \geq 0$ for every x in $[a,b]$, then $\int_a^b f(x)dx \geq 0$

Proof

From the previous section, Area A of the region under the graph of $f(x) \geq 0$ from a to $b (a < b)$ is $A = \int_a^b f(x)dx$

Area is Nonnegative! $A = \int_a^b f(x)dx \geq 0$

Corollary

If f and g are integrable on $[a,b]$ and $f(x) \geq g(x)$ for every x in $[a,b]$, then $\boxed{\int_a^b f(x)dx \geq \int_a^b g(x)dx}$

Proof

If f and g are integrable, then $f - g$ is integrable.
$f(x) \geq g(x) \Rightarrow f(x) - g(x) \geq 0$.

From Theorem Above, $\int_a^b [f(x) - g(x)]dx \geq 0$

$\Rightarrow \int_a^b f(x)dx - \int_a^b g(x)dx \geq 0 \Rightarrow \int_a^b f(x)dx \geq \int_a^b g(x)dx$

Example 3

Show that:

$$\int_{-1}^2 \left(\frac{1}{2}x^2 + 3\right)dx \geq \int_{-1}^2 (2x-1)dx$$

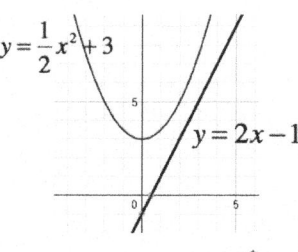

By the corollary,

since $\frac{1}{2}x^2 + 3 \geq 2x - 1$,

$$\boxed{\int_{-1}^2 \left(\frac{1}{2}x^2 + 3\right)dx \geq \int_{-1}^2 (2x-1)dx}$$

Mean Value Theorem for Definite Integrals. Average value of a Function.

If f is continuous on a closed interval $[a,b]$, then there is a number z in the open interval (a,b) such that $\int_a^b f(x)dx = f(z)(b-a)$

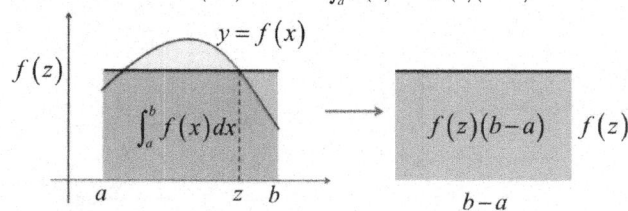

$\int_a^b f(x)dx = f(z)(b-a) \rightarrow f(z) = f_{av}$

If f is continuous on a closed interval $[a,b]$, then the average value of f on $[a,b]$ is $f_{av} = \frac{1}{b-a}\int_a^b f(x)dx$

Average Value of a Function Definition

Let f be continuous on $[a,b]$. The average value f_{av} of f on $[a,b]$ is $\boxed{f_{av} = \frac{1}{b-a}\int_a^b f(x)dx}$

Example 4

Find the average value of $f(x) = x^3$ on $[0,4]$ if it is given that $\int_0^4 x^3 dx = 64$.

By definition, with $a = 0, b = 4$ and $f(x) = x^3$,

$$f_{av} = \frac{1}{4-0}\int_0^4 x^3 dx = \frac{1}{4}(64) = \boxed{16}$$

4.7 First Fundamental Theorem of Calculus

 Mathboat.com

The First Fundamental Theorem of Calculus

Theorem. Let f be continuous on $[a,b]$. Then $\int_a^x f(t)dt$ is differentiable at every point of $[a,b]$ and $\dfrac{d}{dx}\int_a^x f(t)dt = f(x)$

Proof. Let function $G(x)$ be the area under the graph of $f(x)$ from a to x, or
$G(x) = \int_a^x f(t)dt$. Let's prove $G'(x) = f(x)$
Area of one thin Strip approximately is:
$G(x+h) - G(x) = f(\bar{x}) \cdot h$
$\dfrac{G(x+h) - G(x)}{h} = f(\bar{x})$
As $h \to 0$, $\bar{x} \to x$, therefore
$G'(x) = \lim\limits_{h \to 0} \dfrac{G(x+h) - G(x)}{h} = \lim\limits_{\bar{x} \to x} f(\bar{x}) = f(x)$
and so $G'(x) = f(x)$

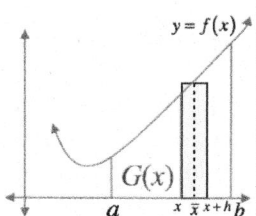

$\boxed{\dfrac{d}{dx}\int_a^x f(t)dt = f(x)}$

Example 1

$\boxed{\dfrac{d}{dx}\int_a^x f(t)dt = f(x)}$

$\dfrac{d}{dx}\int_{-\pi}^{x} \cos t \, dt = \boxed{\cos x}$

By 1st Fundamental Theorem of Calculus

Example 2

$\dfrac{d}{dx}\int_0^x \dfrac{\sin t}{t^2+1}dt = \dfrac{\sin x}{x^2+1}$

By 1st Fundamental Theorem of Calculus

Example 3 (use calculators)

The slope of the function $f(x) = \int_0^x (\arcsin t)dt$ when $x = 0.4$ is

A. 0.081
B. 0.389
C. 0.412
D. 1.091
E. 2.000

By 1st Fundamental Theorem of Calculus,

$\dfrac{d\int_0^x (\arcsin t)dt}{dx} = \arcsin x$

$\arcsin x \big|_{x=0.4} \approx 0.412$

Example 4

$\boxed{\dfrac{d}{dx}\int_a^x f(t)dt = f(x)}$

By 1st Fundamental Theorem of Calculus and Chain Rule,

$\dfrac{d}{dx}\int_0^{x^2} \cos t \, dt = \cos x^2 \cdot (x^2)' = \boxed{2x \cos x^2}$

Example 5

From Example 4

$\dfrac{d}{dx}\int_{x^2}^0 \cos t \, dt = \dfrac{d}{dx}\left(-\int_0^{x^2}\cos t \, dt\right) = \boxed{-2x\cos x^2}$

$\int_a^c f(x)dx = -\int_c^a f(x)dx$

$\dfrac{d}{dx}\int_{\cos^2 5x}^{2} \dfrac{3}{1+t^2}dt = -\dfrac{d}{dx}\int_2^{\cos^2 5x} \dfrac{3}{1+t^2}dt$

$= -\dfrac{3}{1+(\cos^2 5x)^2} \cdot (\cos^2 5x)'$

$= -\dfrac{3}{1+(\cos^2 5x)^2} \cdot 2\cos 5x \cdot (\cos 5x)'$

$= -\dfrac{3}{1+(\cos^2 5x)^2} \cdot 2\cos 5x \cdot -\sin 5x \cdot (5x)'$

$= -\dfrac{3}{1+(\cos^2 5x)^2} \cdot 2\cos 5x \cdot -\sin 5x \cdot 5$

$= \dfrac{15 \cdot (2\cos 5x \sin 5x)}{1+(\cos^2 5x)^2} = \boxed{\dfrac{15 \sin 10x}{1+\cos^4 5x}}$

Example 6

Flip the limits of integration and change the sign

By 1st Fundamental Theorem of Calculus and Chain Rule

4.8a Second Fundamental Theorem of Calculus

 Mathboat.com

Second Fundamental Theorem of Calculus

If f is continuous on [a, b] and F is any antiderivative of f then

$$\int_a^b f(t)\,dt = F(b) - F(a)$$

$$\text{or:} \int_a^b f'(t)\,dt = f(b) - f(a)$$

⇑

derivative or rate of change of f

Second Fundamental Theorem of Calculus

$$\int_a^b f(t)\,dt = F(b) - F(a)$$

Proof

Let $G(x) = \int_a^x f(t)\,dt$

By 1st Fundamental Theorem of Calculus:

$$\frac{d}{dx}\int_a^x f(t)\,dt = f(x)$$

Then $G'(x) = f(x) \Rightarrow G(x) = F(x) + C$

Plug $x = a$ into $G(x)$

Second Fundamental Theorem of Calculus (Proof Continued)

$x = a \Rightarrow G(a) = \int_a^a f(t)\,dt = 0 \Rightarrow 0 = F(a) + C$

$C = -F(a)$

$$\int_a^x f(t)\,dt = F(x) + C \rightarrow \int_a^x f(t)\,dt = F(x) - F(a)$$

Replace x with b

$$\int_a^b f(t)\,dt = F(b) - F(a)$$

Example 1

Evaluate

$$\int_1^2 (4x^{-5} - 3x^2)\,dx = \left(4 \cdot \frac{x^{-4}}{-4} - 3 \cdot \frac{x^3}{3}\right)\bigg|_1^2$$ — Use 2nd Fundamental Theorem of Calculus

$= (-x^{-4} - x^3)\big|_1^2 = -(x^{-4} + x^3)\big|_1^2$ — Simplify the Antiderivative

$= -\left[(2^{-4} + 2^3) - (1^{-4} + 1^3)\right]$ — Plug in limits of integration

$= -\left(\frac{1}{16} + 8 - 2\right) = \boxed{-6\frac{1}{16}}$

Example 2

Evaluate $\int_6^{12} \frac{7}{\sqrt{4x+1}}\,dx$

Use Substitution:
$$\begin{array}{l} u = 4x+1 \\ du = 4\,dx \\ dx = \dfrac{du}{4} \end{array}$$

$\Rightarrow \dfrac{7}{4}\int \dfrac{du}{\sqrt{u}}$ — Substitute. Since u is not changing from 6 to 12, do not write any limits of integration

$= \dfrac{7}{4} \cdot \dfrac{u^{\frac{1}{2}}}{\frac{1}{2}}$ — Integrate. Still do not write any limits of integration

$\Rightarrow \dfrac{7}{2}\left[\sqrt{4x+1}\right]_6^{12}$ — Plug u = 4x+1 back in. Now you should write the limits of integration

$= \dfrac{7}{2}\left(\sqrt{4 \cdot 12 + 1} - \sqrt{4 \cdot 6 + 1}\right) = \dfrac{7}{2}(7 - 5) = \boxed{7}$ — Plug in x = 6 and x = 12

Example 2 (Using Substitution and Changing the Limits of Integration)

This time, we will change the Limits of Integration for variable u

Evaluate

$$\int_{6}^{12} \frac{7}{\sqrt{4x+1}}dx = \frac{7}{4}\int_{25}^{49}\frac{du}{\sqrt{u}} = \frac{7}{4}\left[\frac{u^{\frac{1}{2}}}{\frac{1}{2}}\right]_{25}^{49}$$

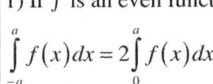
To find the new limits of integration, we plug original ones (x=12 and x=6) into u=4x+1

$x = 12 \Rightarrow u = 4 \cdot 12 + 1 = 49$
$x = 6 \Rightarrow u = 4 \cdot 6 + 1 = 25$

$$= \frac{7}{2}u^{\frac{1}{2}}\Big|_{25}^{49} = \frac{7}{2}\left(49^{\frac{1}{2}} - 25^{\frac{1}{2}}\right)$$

$$= \frac{7}{2}(7-5) = \boxed{7}$$

Theorem

Let f be continuous on $[-a, a]$.

1) If f is an even function,

$$\int_{-a}^{a} f(x)dx = 2\int_{0}^{a} f(x)dx$$

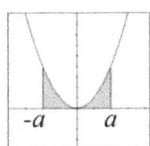

Intuition: Since the regions under the graph on the left and right sides of y-axis have the same area, we can just double the area on one side

2) If f is an odd function,

$$\int_{-a}^{a} f(x)dx = 0$$

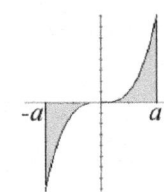

Intuition: The regions still have the same area but the corresponding integrals will have opposite signs since the left region is below the x-axis and the right one is above. We can just cancel these integrals to get 0

Example 3

Evaluate $\int_{-\frac{\pi}{6}}^{\frac{\pi}{6}} (x + \sin 4x) dx$

Since the limits of integration are opposite numbers, check if the integrand is odd or even function

$f(x) = x + \sin 4x$ $\therefore f(x)$ is Odd Function
$f(-x) = -x + \sin(-4x) =$
$-(x + \sin 4x) =$
$-x - \sin 4x = -f(x)$ $\therefore \int_{-\frac{\pi}{6}}^{\frac{\pi}{6}} (x + \sin 4x) dx = \boxed{0}$

Example 4

Evaluate $\int_{-1}^{1}(5x^4 + 3x^2 - 1)dx$

Since the limits of integration are opposite numbers, check if the integrand is odd or even function

$f(x) = 5x^4 + 3x^2 - 1$
$f(-x) = 5(-x)^4 + 3(-x)^2 - 1 = 5x^4 + 3x^2 - 1 = f(x)$
$\therefore f(x)$ is Even Function

$$\therefore \int_{-a}^{a} f(x)dx = 2\int_{0}^{a} f(x)dx$$

Example 4 Continued

$$\int_{-1}^{1}(5x^4 + 3x^2 - 1)dx =$$

$$2\int_{0}^{1}(5x^4 + 3x^2 - 1)dx =$$

$$2\left[\frac{5x^5}{5} + \frac{3x^3}{3} - x\right]_{0}^{1}$$

$$= 2\left[x^5 + x^3 - x\right]_{0}^{1} =$$

$$2(1 + 1 - 1) = \boxed{2}$$

Example 5

A pizza, heated to a temperature of 350 degrees Fahrenheit (°F), is taken out of an oven and placed in a 75°F room at time t = 0 minutes. The temperature of the pizza is changing at a rate of $-110e^{-0.4t}$ degrees Fahrenheit per minute. To the nearest degree, what is the temperature of the pizza at time t = 5 minutes?

Solution

$$\int_{0}^{5} -110e^{-.4t}dt = T(5) - T(0)$$

Use graphing calculator to integrate:

$-238 = T(5) - 350$

$T(5) = 350 - 238 = \boxed{112}$

4.8 b
Indeterminate Forms.
L'Hôpital's (L'Hospital's) Rule.

Note: For "Late Transcendental" approach, read examples 3, 4 and 5 at the end of Chapter 6.

 Mathboat.com

Indeterminate Forms

While solving limits, such as: $\lim_{x \to c} \dfrac{f(x)}{g(x)}$

you may come up with:

$\lim_{x \to c} f(x) = 0$ which leads to

$\lim_{x \to c} g(x) = 0$ indeterminate form: $\dfrac{0}{0}$

Or:

$\lim_{x \to c} f(x) = \infty$ which leads to

$\lim_{x \to c} g(x) = \infty$ indeterminate form: $\dfrac{\infty}{\infty}$

Use L'Hospital's Rule!   What is L'Hospital's Rule?

L'Hôpital's (L'Hospital's) Rule

Suppose f and g are differentiable on an open interval (a,b) containing c, except possibly at c itself.

If $\dfrac{f(x)}{g(x)}$ has the indeterminate form $\dfrac{0}{0}$ or $\dfrac{\infty}{\infty}$ at $x = c$ and if $g'(x) \neq 0$ for $x \neq c$,

then $\lim_{x \to c} \dfrac{f(x)}{g(x)} = \lim_{x \to c} \dfrac{f'(x)}{g'(x)}$, provided either

$\lim_{x \to c} \dfrac{f'(x)}{g'(x)}$ exists or $\lim_{x \to c} \dfrac{f'(x)}{g'(x)} = \infty$

First Form of L'Hôpital's Rule

$$f(a) = g(a) = 0, \quad g'(a) \neq 0 \Rightarrow \lim_{x \to a} \dfrac{f(x)}{g(x)} = \dfrac{f'(a)}{g'(a)}$$

Proof. Given that $f(a) = g(a) = 0$.

By Definition of Derivative,

$$\dfrac{f'(a)}{g'(a)} = \dfrac{\lim_{x \to a} \dfrac{f(x) - \cancel{f(a)}}{\cancel{x - a}}}{\lim_{x \to a} \dfrac{g(x) - \cancel{g(a)}}{\cancel{x - a}}} = \lim_{x \to a} \dfrac{f(x)}{g(x)}$$

Stronger Form of L'Hôpital's Rule:

$\lim_{x \to a} \dfrac{f(x)}{g(x)} = \lim_{x \to a} \dfrac{f'(x)}{g'(x)}$

(Proof is based on the special version of the Mean Value Theorem)

The proof for the indeterminate form $\dfrac{\infty}{\infty}$ requires more advanced level of Calculus

Example 1

Find $\lim_{x \to 0} \dfrac{2\cos x + 5x - 2}{4x}$

$\boxed{\lim_{x \to 0}(2\cos x + 5x - 2) = 0 \text{ and } \lim_{x \to 0}(4x) = 0}$

Using L'Hospital's Rule,

$\lim_{x \to 0} \dfrac{2\cos x + 5x - 2}{4x} = \lim_{x \to 0} \dfrac{(2\cos x + 5x - 2)'}{(4x)'} = \lim_{x \to 0} \dfrac{-2\sin x + 5}{4}$

$= \dfrac{0 + 5}{4} = \boxed{\dfrac{5}{4}}$

Example 2. $\lim\limits_{x \to 4} \dfrac{x^2 - 16}{\int_4^x \cos(\pi t)\, dt} = ?$

$\lim\limits_{x \to 4}(x^2 - 16) = 0$ and $\lim\limits_{x \to 4}\int_4^x \cos(\pi t)\, dt = 0$

Using L'Hospital's Rule,

$\lim\limits_{x \to 4} \dfrac{x^2 - 16}{\int_4^x \cos(\pi t)\, dt} = \lim\limits_{x \to 4} \dfrac{(x^2 - 16)'}{\dfrac{d\int_4^x \cos(\pi t)\, dt}{dx}}$

By 1st Fundamental Theorem of Calculus,

$= \lim\limits_{x \to 4} \dfrac{2x}{\cos(\pi x)} = \dfrac{2 \cdot 4}{\cos 4\pi} = \dfrac{8}{1} = \boxed{8}$

Example 3 (for Early Transcendentals)

Find $\lim\limits_{x \to 0} \dfrac{e^x + e^{-x} - 2}{1 - \cos 4x}$

$\lim\limits_{x \to 0}(e^x + e^{-x} - 2) = 1 + 1 - 2 = 0$ and

$\lim\limits_{x \to 0}(1 - \cos 4x) = 1 - 1 = 0$

Using L'Hospital's Rule,

$\lim\limits_{x \to 0} \dfrac{e^x + e^{-x} - 2}{1 - \cos 4x} = \lim\limits_{x \to 0} \dfrac{(e^x + e^{-x} - 2)'}{(1 - \cos 4x)'} = \lim\limits_{x \to 0} \dfrac{e^x - e^{-x}}{4 \sin 4x}$

Example 3 Continued

$\lim\limits_{x \to 0}(e^x - e^{-x}) = 0$ and $\lim\limits_{x \to 0}(4 \sin 4x) = 0$

Using L'Hospital's Rule Again,

$\lim\limits_{x \to 0} \dfrac{e^x - e^{-x}}{4 \sin 4x} = \lim\limits_{x \to 0} \dfrac{(e^x - e^{-x})'}{(4 \sin 4x)'} =$

$\lim\limits_{x \to 0} \dfrac{e^x + e^{-x}}{16 \cos 4x} = \dfrac{1 + 1}{16 \cos 0} = \dfrac{2}{16} = \boxed{\dfrac{1}{8}}$

Example 4 (for Early Transcendentals)

Find $\lim\limits_{x \to \infty} \dfrac{\ln x + 3}{\sqrt{x}}$

$\lim\limits_{x \to \infty}(\ln x + 3) = \infty$ and $\lim\limits_{x \to \infty}(\sqrt{x}) = \infty$

Using L'Hospital's Rule,

$\lim\limits_{x \to \infty} \dfrac{\ln x + 3}{\sqrt{x}} = \lim\limits_{x \to \infty} \dfrac{(\ln x + 3)'}{(\sqrt{x})'} = \lim\limits_{x \to \infty} \dfrac{\dfrac{1}{x}}{\dfrac{1}{2\sqrt{x}}}$

$= \lim\limits_{x \to \infty} \dfrac{2\sqrt{x}}{x} = \lim\limits_{x \to \infty} \dfrac{2}{\sqrt{x}} = \boxed{0}$

Example 5 (for Early Transcendentals)

Find $\lim\limits_{x \to \infty} \dfrac{e^{4x} - 2}{x^2}$

$\lim\limits_{x \to \infty}(e^{4x} - 2) = \infty$ and $\lim\limits_{x \to \infty}(x^2) = \infty$

Using L'Hospital's Rule,

$\lim\limits_{x \to \infty} \dfrac{e^{4x} - 2}{x^2} = \lim\limits_{x \to \infty} \dfrac{(e^{4x} - 2)'}{(x^2)'} = \lim\limits_{x \to \infty} \dfrac{4e^{4x}}{2x} = \lim\limits_{x \to \infty} \dfrac{2e^{4x}}{x}$

Example 5 Continued

$\lim\limits_{x \to \infty}(2e^{4x}) = \infty$ and $\lim\limits_{x \to \infty}(x) = \infty$

Using L'Hospital's Rule Again,

$\lim\limits_{x \to \infty} \dfrac{2e^{4x}}{x} = \lim\limits_{x \to \infty} \dfrac{(2e^{4x})'}{(x)'} = \lim\limits_{x \to \infty} \dfrac{8e^{4x}}{1} = \boxed{\infty}$

4.9 Numerical Integration. Trapezoidal Approximation.

Trapezoidal Approximation

With the Trapezoidal Approximation, instead of approximating area by using rectangles (as you do with the left, right, and midpoint Riemann sum methods), you approximate area with trapezoids.

Example 1

The values of a differentiable function $G(t)$ at several points are shown in the table below. Using trapezoidal sum with four subintervals, what is the area under the curve $G(t)$?

t	0	3	5	7	10
$G(t)$	0	5	11	20	22

Area of Trapezoid = $\dfrac{base_1 + base_2}{2} \cdot height$

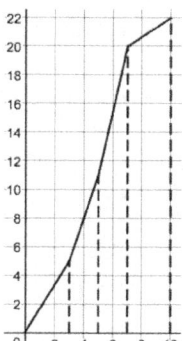

Area under the curve $G(t)$: $A = \int_a^b G(t)\,dt$

$$A = \int_0^{10} G(t)\,dt \approx \frac{1}{2}\cdot(3-0)\cdot(0+5) + \frac{1}{2}(5-3)\cdot(5+11)$$
$$+\frac{1}{2}(7-5)\cdot(11+20) + \frac{1}{2}(10-7)\cdot(20+22) \approx 117.5$$

Example 2

The velocity of a particle moving along a straight line is modeled by a differentiable function v. Selected values of $v(t)$, in meters/seconds, are given in the table below. If 5 subintervals are used, what is the trapezoidal approximation of $v(t)$ from time $t = 20$ to $t = 40$?

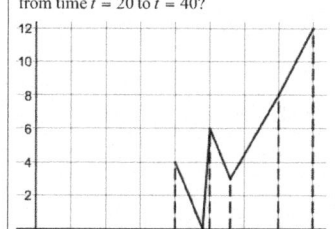

$t_{(sec)}$	20	24	25	28	35	40
$v(t)$	4	0	6	3	8	12

Area of Trapezoid = $\dfrac{base_1 + base_2}{2} \cdot height$

$$A = \int_{20}^{40} V(t)\,dt \approx \frac{1}{2}\cdot(24-20)\cdot(0+4) + \frac{1}{2}(25-24)\cdot(6+0)+$$
$$\frac{1}{2}(28-25)\cdot(3+6) + \frac{1}{2}(35-28)\cdot(8+3) + \frac{1}{2}(40-35)\cdot(12+8) \approx 113$$

Example 3

The temperature, in degrees Celsius, of the water in a lake is a differentiable function W of time t. The table below shows the water temperature over a 15-day period. Find the average temperature, in degrees Celsius, of the water over the time interval $0 \le t \le 15$ days, using a trapezoidal approximation.

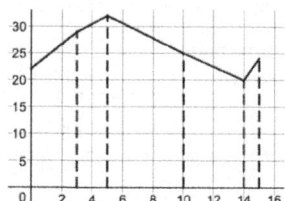

$t_{(days)}$	$W(t)$
0	22
3	29
5	32
10	25
14	20
15	24

Example 3 Continued

Area of Trapezoid = $\dfrac{base_1 + base_2}{2} \cdot height$

$$\text{Average temperature} = \frac{1}{15-0}\int_0^{15} W(t)\,dt$$

$$\approx \frac{1}{15}\cdot\left[\frac{1}{2}\cdot(3-0)\cdot(29+22) + \frac{1}{2}(5-3)\cdot(32+29)+\right.$$
$$\frac{1}{2}(10-5)\cdot(25+32) + \frac{1}{2}(14-10)\cdot(20+25)+$$
$$\left.\frac{1}{2}(15-14)\cdot(24+20)\right] = \frac{1}{15}\cdot 392 \approx \boxed{26.1°C}$$

Example 4

A train runs back and forth on a south-north section of railroad track. Train's velocity, measured in meters per minute, is given by a differentiable function $v(t)$, where time t is measured in minutes. Selected values for $v(t)$ are given in the table below.

t (minutes)	0	3	6	8	10
$v(t)$ (meters/min)	0	90	30	-110	-130

At time $t = 3$, train's position is 400 meters south of the Central Station, and the train is moving to the south. Write an expression involving an integral that gives the position of the train in meters from the Central Station, at time $t = 10$. Use a trapezoidal sum with three subintervals indicated by the table to approximate the position of the train at time $t = 10$.

Example 4 Solution

$$\int_3^{10} v(t)\,dt = x(10) - x(3)$$

$$x(10) = x(3) + \int_3^{10} v(t)\,dt = 400 + \int_3^{10} v(t)\,dt$$

Area of Trapezoid = $\frac{1}{2} h(b_1 + b_2)$

Using a trapezoidal approximation for the definite integral:

$$x(10) = 400 + \frac{1}{2}(6-3)(90+30) + \frac{1}{2}(8-6)(30-110)$$
$$+ \frac{1}{2}(10-8)(-110+-130) = \boxed{260}$$

Trapezoidal Rule (not tested on the AP Exam)

Let f be continuous on $[a,b]$ and regular partition of $[a,b]$ is determined by $a = a_0, a_1, ..., a_n = b$

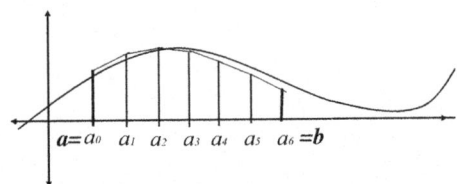

$a = a_0 \quad a_1 \quad a_2 \quad a_3 \quad a_4 \quad a_5 \quad a_6 = b$

Trapezoidal Rule gives the **sum of the areas of the trapezoids** under the curve

$$\int_a^b f(x)\,dx = \frac{b-a}{2n}\left[f(a_0) + 2f(a_1) + 2f(a_2) + ... + 2f(a_{n-1}) + f(a_n)\right]$$

Proof of the Trapezoidal Rule

Let f be continuous on $[a,b]$ and regular partition of $[a,b]$ is determined by $a = a_0, a_1, ..., a_n = b$

$$\int_a^b f(x)\,dx = \frac{b-a}{2n}\left[f(a_0) + 2f(a_1) + 2f(a_2) + ... + 2f(a_{n-1}) + f(a_n)\right]$$

1. Area of each individual trapezoid: $\frac{\text{base}_1 + \text{base}_2}{2} \cdot \text{height}$

From the diagram,
$b_1 = f(a_0)$
$b_2 = f(a_1)$
$h = a_1 - a_0$

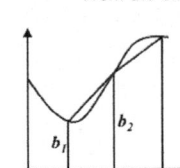

Thus, area of one trapezoid is

$$\frac{f(a_0) + f(a_1)}{2} \cdot (a_1 - a_0)$$

Proof of the Trapezoidal Rule Continued

2. Since we divide the curve equally in the x-direction, and since there are in total of n blocks, $a_1 - a_0 = ... = a_n - a_{n-1} = \frac{b-a}{n}$

Let's find and **add the areas of 3 trapezoids**:

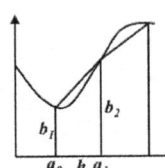

$\frac{1}{2} \cdot \frac{b-a}{n} \cdot [f(a_0) + f(a_1)]$
$\frac{1}{2} \cdot \frac{b-a}{n} \cdot [f(a_1) + f(a_2)]$
$\frac{1}{2} \cdot \frac{b-a}{n} \cdot [f(a_2) + f(a_3)]$

Yielding the form: $\frac{1}{2} \cdot \frac{b-a}{n} \cdot [f(a_0) + 2f(a_1) + 2f(a_2) + f(a_3)]$

$$\int_a^b f(x)\,dx = \frac{b-a}{2n}\left[f(a_0) + 2f(a_1) + 2f(a_2) + ... + 2f(a_{n-1}) + f(a_n)\right]$$

Example 5

The values of the function f(x) are given in the table below:

x	0	1	2	3	4	5	6	7	8	9	10
f(x)	20	19.5	18	15.5	12	7.5	2	-4.5	-12	-20.5	-30

Use the Trapezoidal Rule to approximate:

$$\int_a^b f(x)\,dx \approx \frac{b-a}{2n} \cdot \left[f(a_0) + 2f(a_1) + 2f(a_2) + ... + 2f(a_{n-1}) + f(a_n)\right]$$

$$\int_0^{10} f(x)\,dx \approx \frac{10-0}{2(10)}[20 + 2(19.5) + 2(18) + 2(15.5) + 2(12) + 2(7.5) +$$
$$2(2) + 2(-4.5) + 2(-12) + 2(-20.5) + (-30)] = 32.5$$

4.10 Slope Fields

Mathboat.com

Determining Slope Fields Graphs

Point	$\frac{dy}{dx}=-\frac{x}{y}$
(1,1)	$-1/1=-1$
(1,-1)	$-1/-1=1$
(2,1)	$-2/1=-2$
(-2,1)	$--2/1=2$
(1,2)	$-1/2$
(-1,2)	$1/2$
(1,-2)	$1/2$
(-1,-2)	$-1/2$

$$\frac{dy}{dx}=-\frac{x}{y}$$

Plug in numbers for x and y to find the slope at that point.

At every point on the y-axis (x = 0): the slope = 0, tangent is horizontal

At every point on the x-axis (y = 0): the slope doesn't exist, tangent is vertical.
Continue plotting more points and discover that the graph is a family of semicircles.

Tips for Differential Equations and Slope Fields

$\frac{dy}{dx}=0 \Leftrightarrow$ Horizontal Dashes ———

$\frac{dy}{dx}>0 \Leftrightarrow$ Forward Slashes ⁄

$\frac{dy}{dx}<0 \Leftrightarrow$ Back Slashes ＼

$\frac{dy}{dx}$ does not have y
$\Leftrightarrow$ All Dashes are Parallel on Vertical Axis

$\frac{dy}{dx}$ does not have x
$\Leftrightarrow$ All Dashes are Parallel on Horizontal Axis

$\frac{dy}{dx}$ is Periodic
$\Leftrightarrow$ Dashes look like a periodic function

Describe the Slope Field 1

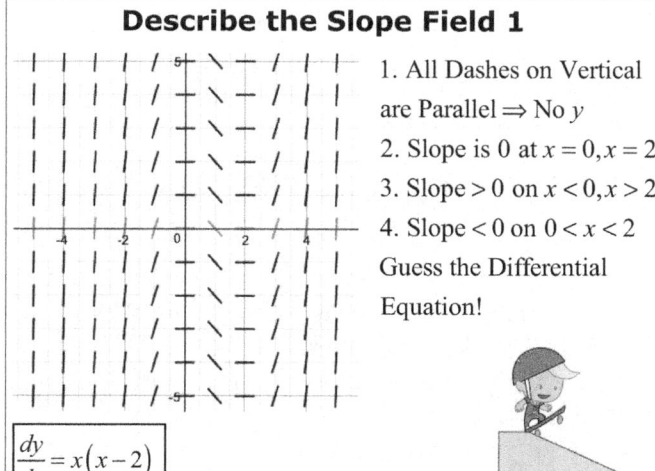

1. All Dashes on Vertical are Parallel $\Rightarrow$ No y
2. Slope is 0 at $x=0, x=2$
3. Slope >0 on $x<0, x>2$
4. Slope <0 on $0<x<2$

Guess the Differential Equation!

$\frac{dy}{dx}=x(x-2)$

Describe the Slope Field 2

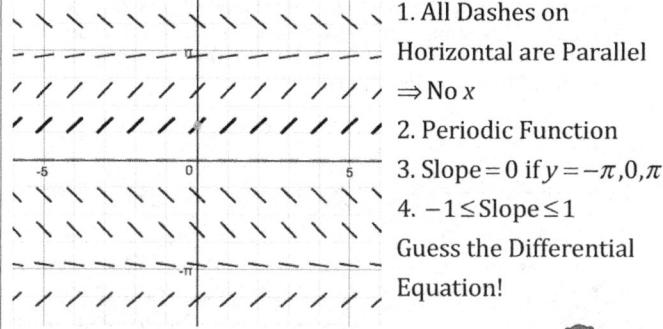

1. All Dashes on Horizontal are Parallel $\Rightarrow$ No x
2. Periodic Function
3. Slope $=0$ if $y=-\pi, 0, \pi$
4. $-1\leq$ Slope ≤ 1

Guess the Differential Equation!

$\frac{dy}{dx}=\sin y$

Identify the Slope Field 1

$\frac{dy}{dx}=2x^2$

A B C

D E F

Identify the Slope Field 1

 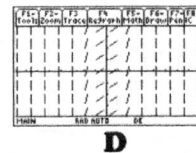

D

$$\frac{dy}{dx} = 2x^2$$

Since 2x² is always positive, the dashes should always be Forward Slashes.
This was the only slope field with such conditions.

Identify the Slope Field 2

Slope should be 0 when y = -3 (field should show Horizontal). This was the only slope field with an asymptote at y = -3.

$$\frac{dy}{dx} = y + 3$$

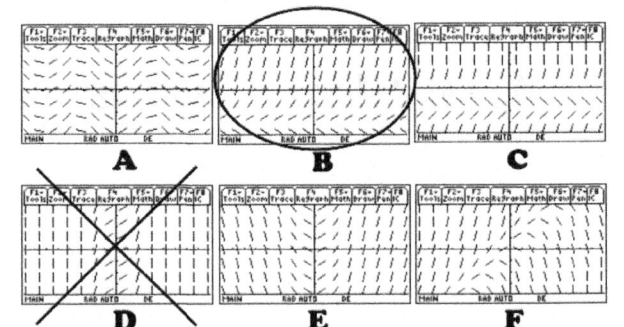

Identify the Slope Field 3

When x=y, the slope = 0, and the dashes should be Horizontal. This was the only slope field to fit such description.

$$\frac{dy}{dx} = y - x$$

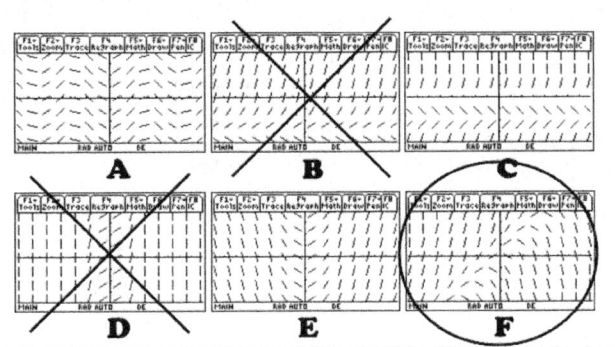

Identify the Slope Field 4

Remember, when there is no x, the lines are Parallel on the Horizontal. Plus there is an asymptote at y = -2. This was the only graph that fit such descriptions.

$$\frac{dy}{dx} = y(y+2)$$

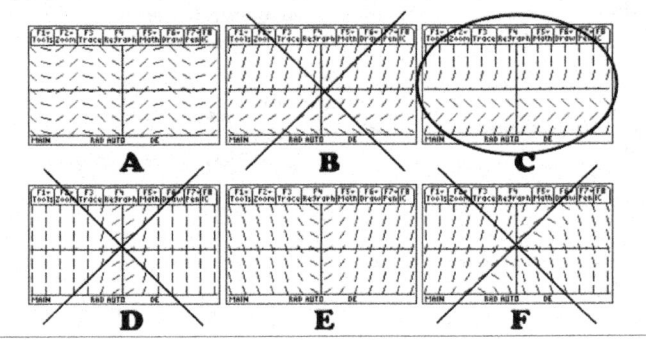

Identify the Slope Field 5

$$\frac{dy}{dx} = x$$

Remember, when there is no y, the lines are Parallel on the vertical. When x>0, dy/dx>0. When absolute value of x is increasing, the dashes become steeper.

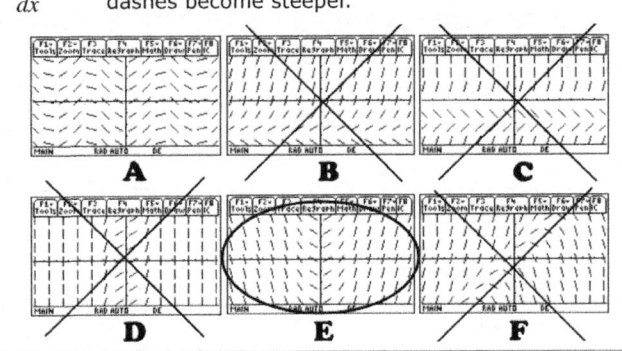

Identify the Differential Equation

Since the slope field is periodic, we usually think of trig equations. This slope field resembles –cos x. Remember that the derivative of –cos x is sin x, so the differential equation will be: **dy/dx = sin x**

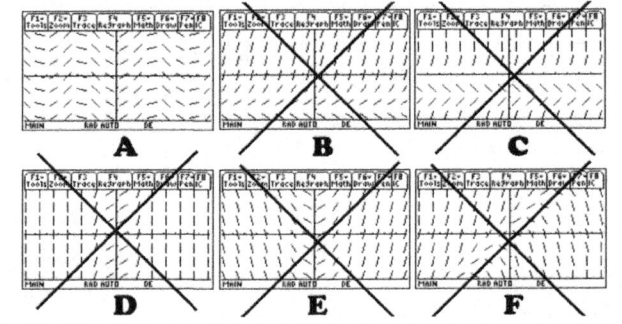

Slope Field on TI-89

Push "Mode"
  For Graph, select "6: Diff Equations"
  Press "Enter" twice to save

Select "", then "F1"
Set tØ = initial x
  (If you don't know, set 0 for now)
y1' = dy/dx
  Instead of x, type in t
  Instead of y, type in y1, y2, or y#
  (# is the number in: y_')

Select "F2"
Select 4:ZoomDec ("F2" + 4") for best graph

Select "", then "F3" to graph

To draw a sample line through a point in slope field
Set tØ = initial x
Set yi1 = initial y

Solution curve. Example 1

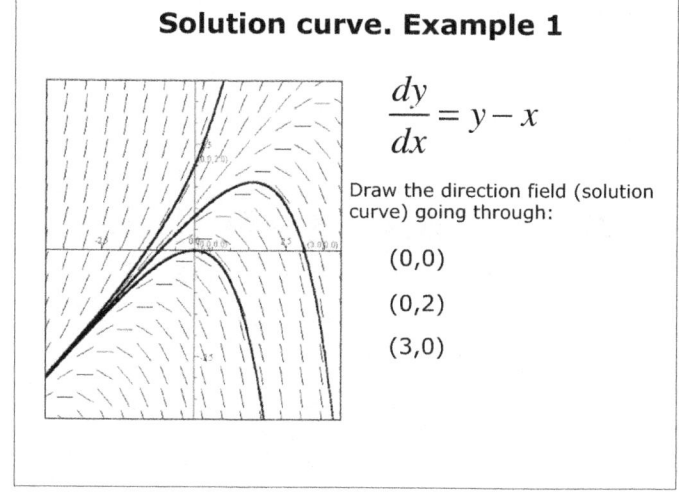

$$\frac{dy}{dx} = y - x$$

Draw the direction field (solution curve) going through:

(0,0)

(0,2)

(3,0)

Solution curve. Example 2

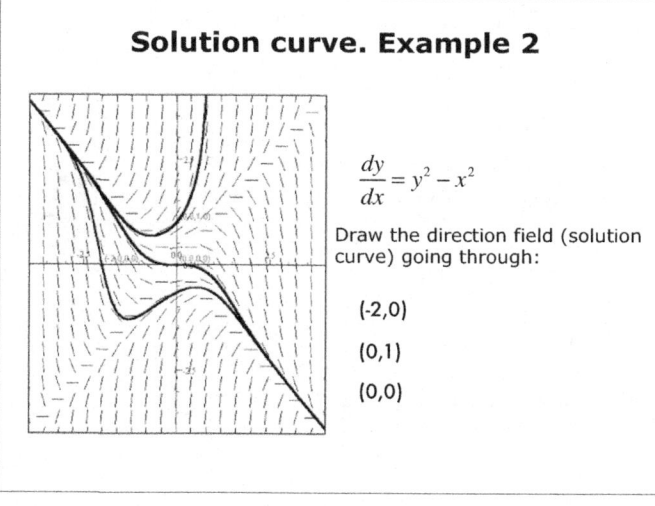

$$\frac{dy}{dx} = y^2 - x^2$$

Draw the direction field (solution curve) going through:

(-2,0)

(0,1)

(0,0)

Solution curve. Example 3

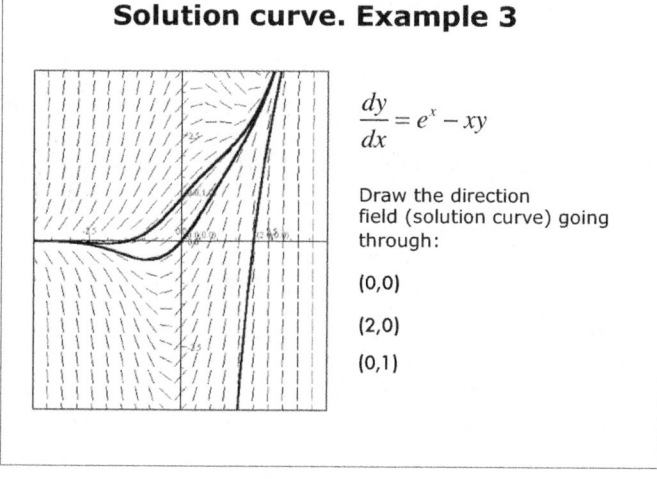

$$\frac{dy}{dx} = e^x - xy$$

Draw the direction field (solution curve) going through:

(0,0)

(2,0)

(0,1)

Example 4

Which of the following is the equation for $\frac{dy}{dx}$ whose field is shown below for $-3 < x < 3$ and $-3 < y < 3$?

$(A)\ \frac{dy}{dx} = y$ $(B)\ \frac{dy}{dx} = x^2 + 1$ $(C)\ \frac{dy}{dx} = \sin x$

$(D)\ \frac{dy}{dx} = x^2 - 9$ $(E)\ \frac{dy}{dx} = 3x^3 - 9x$

(A) All dashes on the same vertical axis are ↑↑ so there is no y involved.

(B) No, since $x^2 + 1$ is always positive, so slopes should all be positive but they are not.

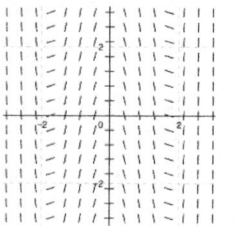

Example 4 Continued

(C) No, because close to $x = -3$ and 3, slopes are really big, so that dashes almost make a vertical line, but $|\sin x| \leq 1$.

(D) No, since $x < |3|$, then $x^2 - 9 < 0$ but the slopes vary from negative to positive.

(E) YES.

$$\frac{dy}{dx} = 3x^3 - 9x = 3x(x^2 - 3) = 0 \Rightarrow x = 0, \pm\sqrt{3}$$

At $x = 0$: slope $= 0$

Around $x = -\sqrt{3}$ and $\sqrt{3}$: slope changes its sign,

so at $x = -\sqrt{3}$ and $\sqrt{3}$: slope $= 0$

4.11 Euler's Method.
Note: Not on the AP Calculus AB Exam

Euler's Method

Approximation of the Solution to a Differential Equation

- Using Tangent Lines
- Making a Chart
- Using Calculator

Using Tangent Lines

Given: $\dfrac{dy}{dx} = 1 + y$ and $y(0) = 0$ and $\Delta x = 0.5$

Construct the approximate y using 3 steps.

Step 1: Construct a Tangent Line at $(0,0)$

Slope: $1 + 0 = 1$

Equation: $y - 0 = 1(x - 0)$

$y = x$

new y: $y(0.5) = 0.5$

Starting Point: $(0,0)$
Ending Point: $(0.5, 0.5)$

Using Tangent Lines (Continued)

Given: $\dfrac{dy}{dx} = 1 + y$ and $y(0) = 0$ and $\Delta x = 0.5$

Construct the approximate y using 3 steps.

Step 2: Construct a Tangent Line at $(0.5, 0.5)$

Slope: $1 + 0.5 = 1.5$

Equation: $y - 0.5 = 1.5(x - 0.5)$

$y = 1.5x - 0.25$

New y: $y(1) = 1.5 - 0.25 = 1.25$

Starting Point: $(0.5, 0.5)$
Ending Point: $(1, 1.25)$

Using Tangent Lines (Continued)

Given: $\dfrac{dy}{dx} = 1 + y$ and $y(0) = 0$ and $\Delta x = 0.5$

Construct the approximate y using 3 steps.

Step 3: Construct a Tangent Line at $(1, 1.25)$

Slope: $1 + 1.25 = 2.25$

Equation: $y - 1.25 = 2.25(x - 1)$

$y = 2.25x - 1$

New y: $y(1.5) = 3.375 - 1 = 2.375$

Starting Point: $(1, 1.25)$
Ending Point: $(1.5, 2.375)$

Now find the exact solution of $\dfrac{dy}{dx} = 1 + y$ with the given information.

Given: $y(0) = 0$

$$\dfrac{dy}{dx} = 1 + y$$

$$\int \dfrac{dy}{1+y} = \int dx$$

$$\ln|1+y| = x + C \Rightarrow C = 0$$

$$\ln|1+y| = x$$

$$|1+y| = e^x$$

$$\boxed{y = e^x - 1}$$

Euler's Method vs Exact Solution

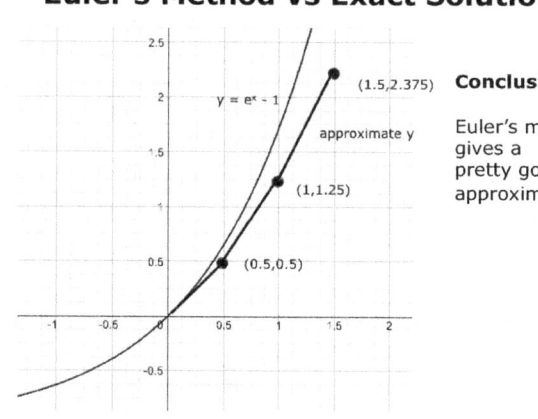

Conclusion:

Euler's method gives a pretty good approximation

Making a Chart

Given: $\dfrac{dy}{dx} = 2x$ and $y(1) = 3$. Find $y(2)$ using 5 steps.

To construct a tangent line, we used the equation of the line: $y - y_0 = m(x - x_0)$

So, the Basic Idea is: New $y \approx$ Old $y + \dfrac{dy}{dx} \Delta x$; Step Size: $\Delta x = \dfrac{(2-1)}{5} = \dfrac{1}{5}$

x	old y	$\Delta y = \dfrac{dy}{dx} \Delta x$	new y = old y + $\dfrac{dy}{dx}\Delta x$
1	3	2(1/5)=0.40	3+0.40=3.40
1.2	3.4	2.4(1/5)=0.48	3.4+0.48=3.88
1.4	3.88	2.8(1/5)=0.56	3.88+0.56=4.44
1.6	4.44	3.2(1/5)=0.64	4.44+0.64=5.08
1.8	5.08	3.6(1/5)=0.72	5.08+0.72=5.80
2.0	5.80		

Using TI-89 Calculator

Given: $\dfrac{dy}{dx} = 2x$ and $y(1) = 3$. $\Delta x = \dfrac{1}{5}$. Find $y(2)$.

Change Mode to "Diff. Equations"

Press ◊ and F1 (Y=) → Enter Initial x Condition under $t0 =$ and Initial y under $yi1 =$

Use t in place of x and $y1$ in place of y when inputting the Differential Equations

Using TI-89 Calculator (Continued)

While still on Y= Page, press F1 and
Scroll Down to 9:
Format, Press Enter
Scroll Down to Solution Method
and choose "Euler"

Press ◊ and F2 to go to the Window Page.
Change the *tstep* to the Δx

Press ◊ and F4 (TblSet) to Input the Initial x under tblStart and the Δx under Δtbl.

Press ◊ and F5 to View the Table

1	3
1.2	3.40
1.4	3.88
1.6	4.44
1.8	5.08
2.0	5.80

Example (using graphing calculator)

Given: $\dfrac{dy}{dx} = -3xy$, where $y(1) = 4$ and $\Delta x = 0.5$.

Approximate $y(2.5)$.

Using TI-89 Calculator,

On Y= Menu:

$t0 = 1$, $y1' = -3t \cdot y1$, $yi1 = 4$

When $x = 2.5$, $\boxed{y \approx -5}$

t	y1
1	4
1.5	-2
2	2.5
2.5	-5

Example (not using graphing calculator)

If $\dfrac{dy}{dx} = xy - y^2$ and $y(2) = 3$, find approximate $y(3)$

using Euler's Method with the step size of 0.5?

x	y	$\Delta y = \dfrac{dy}{dx}\Delta x$	New $y = y + \Delta y$
2	3	$(2 \cdot 3 - 3^2) \cdot (0.5) = -1.5$	$3 + (-1.5) = 1.5$
2.5	1.5	$(2.5 \cdot 1.5 - 1.5^2) \cdot (0.5) = 0.75$	$0.75 + 1.5 = 2.25$
3	2.25		

$$\boxed{y(3) \approx 2.25}$$

5.1 Areas of Plane Regions

Area of Plane Region

Definition

Let the function $f(x)$ be continuous and nonnegative on $[a,b]$.

The Area under the graph of $f(x)$ is $A = \int_a^b f(x)\,dx$

Example 1

Find the area under $f(x) = \sin x$ from $x = 0$ to $x = \dfrac{\pi}{3}$

$$\text{Area} = \int_0^{\frac{\pi}{3}} \sin x\, dx = -\cos x \Big]_0^{\frac{\pi}{3}}$$

$$= -\left[\left(\cos\frac{\pi}{3}\right) - (\cos 0)\right]$$

$$= -\left(\frac{1}{2} - 1\right) = -\left(-\frac{1}{2}\right) = \boxed{\frac{1}{2}}$$

Area between the curves

Theorem

Let the functions f and g be continuous on $[a,b]$ with $f(x) > g(x)$ on $[a,b]$. Then the area between the graphs of f and g from a to b is: $A = \int_a^b (f(x) - g(x))\,dx$

Illustration of the Theorem

If $f(x)$ and $g(x)$ are positive on $[a,b]$:

Area of this shaded region =
Area Under $f(x)$ - Area Under $g(x)$

$$A = \int_a^b (f(x) - g(x))\,dx$$

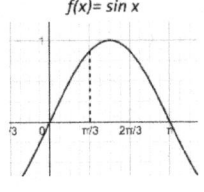

Illustration continued

Let's move $f(x)$ and $g(x)$ 4 units down. The new functions $F(x) = f(x) - 4$ and $G(x) = g(x) - 4$ are not positive on $[a,b]$

But the areas of both regions are still the same.

Area of the 1st region =
Area of the 2nd region =

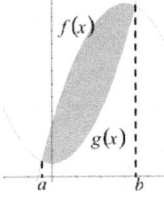

$$\int_a^b (f(x) - g(x))\,dx = \int_a^b ((F(x) + 4) - (G(x) + 4))\,dx$$

$$A = \int_a^b (F(x) - G(x))\,dx$$

Example 2

Find the area of the region bounded by the graphs of $f(x) = 5 - x^2$ and $g(x) = (x-1)^2$

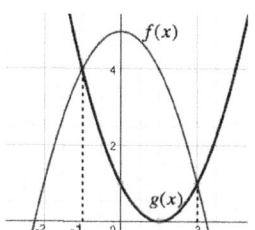

For intersections,
solve: $(x-1)^2 = 5 - x^2$
$x^2 - 2x + 1 = 5 - x^2$
$2x^2 - 2x - 4 = 0$
$2(x+1)(x-2) = 0$
$x = -1, x = 2$

Example 2 Continued

$$\text{Area} = \int_{-1}^{2} (f(x) - g(x))\, dx$$

$$\text{Area} = \int_{-1}^{2} \left((5 - x^2) - (x-1)^2\right) dx$$

$$= \int_{-1}^{2} (4 + 2x - 2x^2)\, dx$$

$$= 4x + x^2 - \frac{2}{3}x^3 \Big]_{-1}^{2}$$

$$= \frac{20}{3} - \frac{-7}{3} = \boxed{9}$$

Example 3

Find the area of the region bounded by the graphs of $f(x) = x^3 + x^2$ and $g(x) = 2x$

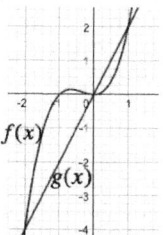

$f(x) > g(x)$ on $[-2, 0]$
$g(x) > f(x)$ on $[0, 1]$

Solve: $x^3 + x^2 = 2x$
$x = \{-2, 0, 1\}$

Example 3 Continued

$$\text{Area} = \int_{-2}^{0} \left[(x^3 + x^2) - (2x)\right] dx + \int_{0}^{1} \left[(2x) - (x^3 + x^2)\right] dx$$

OR :

$$\text{Area} = \left|\int_{-2}^{0} \left[(x^3 + x^2) - (2x)\right] dx\right| + \left|\int_{0}^{1} \left[(2x) - (x^3 + x^2)\right] dx\right|$$

$$= \frac{8}{3} + \frac{5}{12} = \boxed{\frac{37}{12}}$$

Example 4

Find the area of the region bounded by the graphs of $y^2 = x + 1$ and $x - y = 1$.

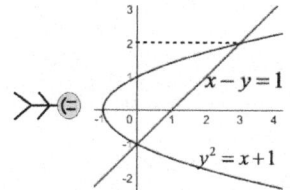

Solve: $y^2 - 1 = y + 1$
$y^2 - y - 2 = 0$
$y = \{-1, 2\}$

If we view graph from perspective of this person, we can write integral in terms of y

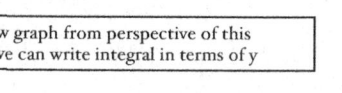

Example 4 Continued

Area =

$$\int_{-1}^{2} \left[(y+1) - (y^2 - 1)\right] dy =$$

$$\int_{-1}^{2} (2 + y - y^2)\, dy =$$

$$2y + \frac{y^2}{2} - \frac{y^3}{3} \Big]_{-1}^{2} =$$

$$\left(4 + 2 - \frac{8}{3}\right) - \left(-2 + \frac{1}{2} + \frac{1}{3}\right) = \boxed{\frac{9}{2}}$$

5.2 The Net Change in Position and Distance Traveled by a Moving Body

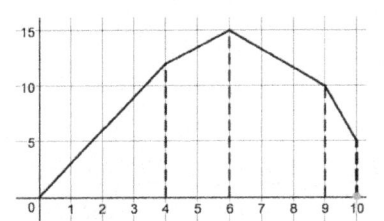

Net Change, Displacement and Total Distance

If a body moving along the line reverses direction while it travels, for example:

It moves 3m forward, then 4m backward, then 6m forward,

then its **Net Change** of the position is: 3−4+6=5m,

and the **Total Distance** is: 3+4+6 = 13m

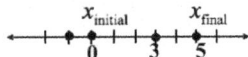

Displacement is defined to be the change in position of an object. It can be defined mathematically with the equation:

$\Delta x = x_{final} - x_{initial} = 5 - 0 = 5$

Net Change in Position or Displacement

If a body moves along the line and its velocity is a continuous function of time, then body's position is:

$$\boxed{s(t) = \int v(t)\,dt = F(t) + C}$$

where F is antiderivative of v.

Net Change in body's Position during time t = a to t = b is

$$\boxed{s(b) - s(a) = \int_a^b v(t)\,dt}$$

Proof
$$s(b) - s(a) = (F(b) + \cancel{C}) - (F(a) + \cancel{C})$$
$$= F(b) - F(a) = \int_a^b v(t)\,dt$$

Example 1

The velocity of a moving body along a line is $v(t) = 5\pi \cos \pi t$ m/sec. Find the net change in the body's position from time $t = 0$ to time $t = 3/2$

$$\int_0^{3/2} 5\pi \cos \pi t \, d\frac{\pi}{\pi} t =$$

$$5 \sin \pi t \Big]_0^{3/2} =$$

$$5(-1 - 0) = \boxed{-5}$$

So, the net effect of the motion from t=0 to t= 3/2 is to shift the body 5 m to the left

Example 2

A particle traveling along the *x*-axis has velocity $v(t) = 8\cos t + 9$. Which of the following is the displacement of the particle over the time interval $\left[0, \frac{\pi}{2}\right]$?

(A) 0   (B) $8 + \frac{9\pi}{2}$ (C) $\frac{9}{2}$ (D) $\frac{9\pi}{2}$

$$\Delta x = \int_{t_1}^{t_2} v(t)\,dt = \int_0^{\frac{\pi}{2}} (8\cos t + 9)\,dt = 8\sin t + 9t \Big]_0^{\frac{\pi}{2}}$$

$$= 8\sin \frac{\pi}{2} + 9 \cdot \frac{\pi}{2} = 8 + \frac{9\pi}{2}$$

Example 3

The table below shows the velocity, in inches per second, at certain times for a particle moving on a line over the time interval $[0,10]$ Which of the following is the displacement of the particle using a trapezoidal sum?

(A) 48   (B) 72   (C) 96   (D) 192

Time t (sec)	0	4	6	9	10
Velocity V(t)(in/sec)	0	12	15	10	5

Displacement = $\int_0^{10} V(t)\,dt \approx$

$\frac{1}{2} \cdot (4-0) \cdot (0+12) +$

$\frac{1}{2}(6-4) \cdot (12+15) +$

$\frac{1}{2}(9-6) \cdot (15+10) +$

$\frac{1}{2}(10-9) \cdot (10+5) = 96$ in

Total Distance Traveled by a Body

If the body moves along the line 6m forward and then 6m backward, the net change of body's position is 0, but the total distance is 12m.

To calculate the total distance with the integral, we have to keep the contributions of the forward motion and a backward motion from canceling each other out.

So integrate the absolute value of the velocity from a to b.

$$\boxed{\text{Total Distance} = \int_a^b |v(t)|\,dt}$$

$$= \left|\int_a^c v(t)\,dt\right| + \left|\int_c^b v(t)\,dt\right|$$

where c is x-intercept of $v(t)$

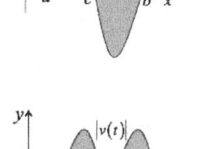

Example 4

The velocity of a moving body along a line is $v(t) = 5\pi \cos \pi t$ m/sec. Find the total distance traveled by the body from time $t = 0$ sec to time $t = 3/2$ sec.

Solution

Find x-intercept of $v(t)$:

$5\pi \cos \pi t = 0$

$\pi t = \dfrac{\pi}{2} + \pi k \Rightarrow t = \dfrac{1}{2} + k$

$k = -1 \Rightarrow t = -\dfrac{1}{2}$ (t can't be < 0);

$k = 0 \Rightarrow \boxed{t = \dfrac{1}{2}}$;

$k = 1 \Rightarrow t = \dfrac{3}{2}$ (endpoint)

Example 4 Continued

Total Distance =

$\int_a^b |v(t)|\,dt = \int_0^{3/2} |5\pi \cos \pi t|\,dt =$

$\left|\int_0^{1/2} 5\pi \cos \pi t\,dt\right| + \left|\int_{1/2}^{3/2} 5\pi \cos \pi t\,dt\right| =$

$5\left(\left|\sin \pi t\Big]_0^{\frac{1}{2}}\right| + \left|\sin \pi t\Big]_{\frac{1}{2}}^{\frac{3}{2}}\right|\right) =$

$5\left(\left|\sin \dfrac{\pi}{2} - \sin 0\right| + \left|\sin \dfrac{3\pi}{2} - \sin \dfrac{\pi}{2}\right|\right)$

$= 5\big[|1-0| + |-1-1|\big] = 5\cdot(1+2) = \boxed{15}$

Example 5

A particle moves along the x-axis such that its acceleration from time $t = 0$ to time $t = 5$ is given by $a(t) = -2\sin t$. The particle has a velocity of 1 at time $t = 0$. What is the total distance traveled by the particle over the time interval?

$\boxed{\text{total distance} = \int_0^5 |v(t)|\,dt}$ $v(t) = \int a(t)\,dt$

$= -2\int \sin t\,dt$

$\boxed{v(t) = 2\cos t + C}$ Plug in: $v(0) = 1$

$2\cos(0) + C = 1$

$2 + C = 1 \Rightarrow C = -1$

$\boxed{v(t) = 2\cos t - 1}$

Example 5 Continued

Total distance $= \int_0^5 |2\cos t - 1|\,dt$

If calculator is allowed, integrate with calculator!!!
(you don't even need to split the integral)

If calculator is not allowed:

Find x-intercepts of $v(t)$: Set $v(t) = 0$

$2\cos t - 1 = 0 \Rightarrow \cos t = \dfrac{1}{2} \Rightarrow \boxed{t = \dfrac{\pi}{3}}$

Example 5 Continued

$v(t) = \left|\int_0^{\frac{\pi}{3}} (2\cos t - 1)\,dt\right| + \left|\int_{\frac{\pi}{3}}^5 (2\cos t - 1)\,dt\right| =$

$\left|2\sin t - t\Big]_0^{\frac{\pi}{3}}\right| + \left|2\sin t - t\Big]_{\frac{\pi}{3}}^5\right| =$

$\left(\left(2\sin \dfrac{\pi}{3} - \dfrac{\pi}{3}\right) - (2\sin 0 - 0)\right) +$

$\left(\left(2\sin \dfrac{\pi}{3} - \dfrac{\pi}{3}\right) - (2\sin 5 - 5)\right) \approx \boxed{8.288}$

5.3 Solids of Revolution. Disk Method.

Mathboat.com

Disk Method

Let's revolve region under $f(x)$ from a to b about x-axis

Separate the area into rectangles of equal length

Revolve each separately to form many disks

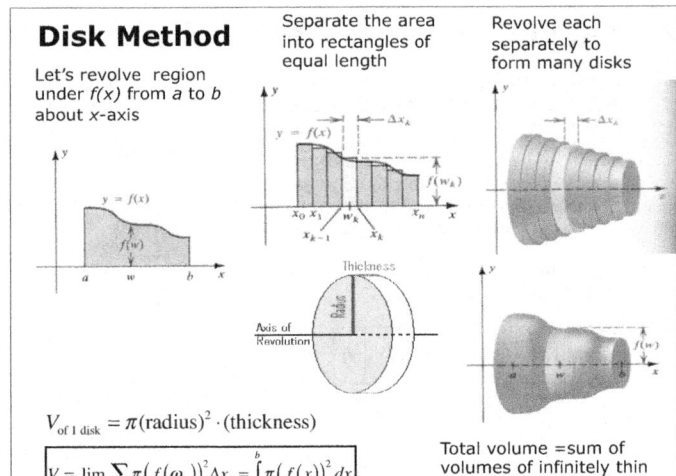

$V_{\text{of 1 disk}} = \pi(\text{radius})^2 \cdot (\text{thickness})$

$$V = \lim_{\|P\|\to 0}\sum_k \pi\big(f(\omega_k)\big)^2 \Delta x_k = \int_a^b \pi\big(f(x)\big)^2 dx$$

Total volume = sum of volumes of infinitely thin individual disks

Formulas

$$V = \pi\int_a^b \big(f(x)\big)^2 dx$$

$$V = \pi\int_c^d \big(g(y)\big)^2 dy$$

Example 1

Find the volume of the solid generated when the region bounded by $y = 3x - x^2$ and the x-axis is revolved around the x-axis.

1) Graph the Region and identify the thickness (dx or dy)

$y = 3x - x^2$

2) Express the radius of one disk in terms of x (if dx) or y (if dy)

$r = 3x - x^2$

3) Integrate. Use Disk method:

$$V = \pi\int_a^b r^2 dx$$

$$V = \pi\int_0^3 (3x - x^2)^2 dx = \pi\int_0^3 (9x^2 - 6x^3 + x^4)dx$$

$$= \pi\left[3x^3 - \frac{3x^4}{2} + \frac{x^5}{5}\right]_0^3 = \pi\left[3\cdot 3^3 - \frac{3\cdot 3^4}{2} + \frac{3^5}{5}\right]$$

Since the region is revolved around the **x-axis**, thickness of disk is dx

$$= \pi\left[81 - \frac{243}{2} + \frac{243}{5}\right] = \boxed{\frac{81\pi}{10}}$$

Example 2

The region bounded by the y-axis and the graphs of $y = \frac{1}{8}x^3$, $y = 0$ and $y = 8$ is revolved about the y-axis. Find the volume of the resulting solid.

Radius of disk $R = \sqrt[3]{8y} = 2y^{\frac{1}{3}}$

Use Disk Method:

$$V = \pi\int_c^d R^2 dy$$

$$V = \pi\int_0^8 \left(2y^{\frac{1}{3}}\right)^2 dy = 4\pi\int_0^8 y^{2/3} dy$$

Since the region is revolved around the **y-axis**, thickness of disk is dy

$$= 4\pi \cdot \frac{3}{5}y^{5/3}\Big|_0^8 = \frac{12\pi}{5}\cdot 8^{\frac{5}{3}} = \frac{12\pi}{5}\cdot(2^3)^{\frac{5}{3}}$$

$$= \frac{12\pi}{5}\cdot 2^5 = \boxed{\frac{384\pi}{5}}$$

Example 3

The region R is bounded by the graph of $y = \sin(2x)$ and $y = 0$ from 0 to π. Find the volume of the solid generated if R is revolved around the x-axis.

Since the regions above and below the x-axis will produce the same volumes of solids of revolution, we can just double one volume

Use Disk Method:

$$V = \pi\int_a^b r^2 dx$$

Thickness of disk: dx
Radius of disk: $\sin 2x$

$$V = 2\int_0^{\frac{\pi}{2}} \pi(\sin^2 2x) dx = 2\int_0^{\frac{\pi}{2}} \pi\left(\frac{1-\cos 4x}{2}\right) dx$$

$$= \pi\left[\int_0^{\frac{\pi}{2}} 1\, dx - \int_0^{\frac{\pi}{2}} \cos 4x\, dx\right] = \pi\left(x\Big|_0^{\frac{\pi}{2}} - \frac{1}{4}\sin(4x)\Big|_0^{\frac{\pi}{2}}\right)$$

$$= \pi\left[\frac{\pi}{2} - \frac{1}{4}(\sin 2\pi - \sin 0)\right] = \frac{1}{2}\pi^2$$

5.4.1 Solids of Revolution. Washer Method.
Part 1. Region is revolved around the *x*-axis and *y*-axis

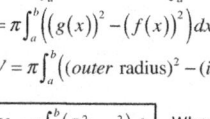 Mathboat.com

Washer Method

Shown at right is the region above the x-axis which is revolved around x-axis. Notice: there is a distance between the region and the axis of revolution

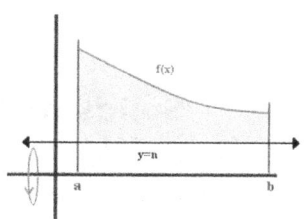

The volume can be imagined to be created from an infinite number of infinitely thin washers. (There is a hole through the solid)

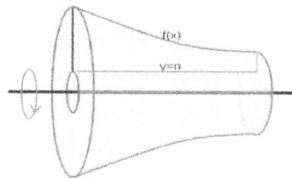

Volume of Solid of Revolution.
Washer Method.

Find the volume V of a solid when the region between the graphs of g(x) and f(x) on [a,b] is rotated around the horizontal line.

V = difference between the volumes created by rotating the regions bounded by the graphs of g(x) and f(x) on [a,b] separately around the same line

$V = \pi \int_a^b (g(x))^2 dx - \pi \int_a^b (f(x))^2 dx$
$= \pi \int_a^b ((g(x))^2 - (f(x))^2) dx$
$V = \pi \int_a^b ((outer\ radius)^2 - (inner\ radius)^2) dx$

R = g(x)
r = f(x)

$\boxed{V = \pi \int_a^b (R^2 - r^2) dx}$ When using Washer or Disc Method, Thickness is *dx* if the region is revolved around the *x*-axis or line || to *x*-axis. Thickness is *dy* if the region is revolved around the *y*-axis or line || to *y*-axis

Example 1

The region bounded by the graphs of $y = \frac{1}{2}x^2 + 3$ and $y = \frac{1}{3}x + 1$ and by the vertical lines $x = -2$ and $x = 3$ is revolved around the *x* - axis. Find the volume of the resulting solid.

Since there is a gap between the region and the axis of revolution, use Washer Method

$V = \pi \int_a^b (R^2 - r^2) dx$

Since the region is revolved around the *x*-axis, the thickness of the washer is *dx*

Outer radius $R = y_{curve} = \frac{1}{2}x^2 + 3$

Inner radius $r = y_{line} = \frac{1}{3}x + 1$

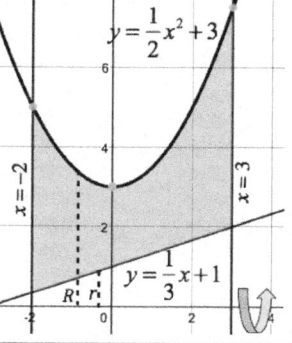

Example 1 Continued

$V = \pi \int_{-2}^{3} \left(\left(\frac{1}{2}x^2 + 3\right)^2 - \left(\frac{1}{3}x + 1\right)^2 \right) dx$

$= \pi \int_{-2}^{3} \left(\frac{1}{4}x^4 + 3x^2 + 9 - \frac{1}{9}x^2 - \frac{2}{3}x - 1 \right) dx$

$= \pi \int_{-2}^{3} \left(\frac{1}{4}x^4 + \frac{26}{9}x^2 - \frac{2}{3}x + 8 \right) dx$

$= \pi \left[\frac{x^5}{20} + \frac{26}{27}x^3 - \frac{x^2}{3} + 8x \right]_{-2}^{3} \ldots \approx \boxed{269.508}$

Example 2

The region in the first quadrant bounded by the graphs of $y = x^2$ and $y = 2x$ is revolved about the *y*-axis. Find the volume of the resulting solid.

Since there is a gap between the region and the axis of revolution, use Washer Method:

$V_{solid} = \pi \int_c^d \left(R_{outer}^2 - r_{inner}^2 \right) dy$

Since the region is revolved around the *y*-axis, the thickness of the washer is *dy*

Outer radius $R = y^{\frac{1}{2}}$, Inner radius $r = \frac{1}{2}y$

$V = \pi \int_0^4 \left[\left(y^{\frac{1}{2}}\right)^2 - \left(\frac{1}{2}y\right)^2 \right] dy$

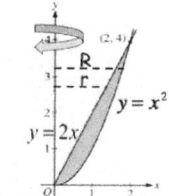

$= \pi \int_0^4 \left(y - \frac{1}{4}y^2 \right) dy = \pi \left(8 - \frac{16}{3} \right) = \boxed{\frac{8}{3}\pi}$

$x^2 = 2x \Rightarrow x = 0, 2$
$y = 0, 4$

5.4.2
Solids of Revolution.
Washer Method
Part 2. Region is revolved around the line parallel to the x-axis and y-axis

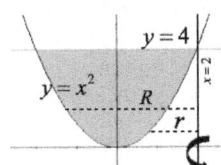 Mathboat.com

Example 1

The region bounded by the graphs of $y = \frac{1}{2}x^2 + 3$ and $y = \frac{1}{3}x + 1$ and by the vertical lines $x = -2$ and $x = 3$ is revolved around the line $y = 8$. Find the volume of the resulting solid.

Since there is a gap between the region and the axis of revolution, use Washer Method:
$$V_{solid} = \pi \int_a^b \left(R_{outer}^2 - r_{inner}^2 \right) dx$$

Since the region is revolved around the line parallel to x-axis, **Thickness : dx**

Outer radius $R = 8 - \left(\frac{1}{3}x + 1 \right) = 7 - \frac{1}{3}x$

Inner radius $r = 8 - \left(\frac{1}{2}x^2 + 3 \right) = 5 - \frac{1}{2}x^2$

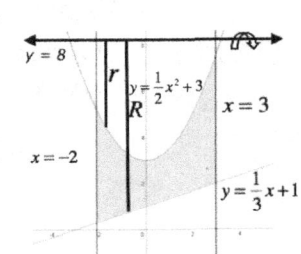

Example 1 Continued

$$V = \pi \int_{-2}^{3} \left(\left(7 - \frac{1}{3}x\right)^2 - \left(5 - \frac{1}{2}x^2\right)^2 \right) dx$$

$$= \pi \int_{-2}^{3} \left(49 - \frac{14}{3}x + \frac{1}{9}x^2 - 25 + 5x^2 - \frac{1}{4}x^4 \right) dx$$

$$= \pi \int_{-2}^{3} \left(-\frac{1}{4}x^4 + \frac{46}{9}x^2 - \frac{14}{3}x + 24 \right) dx$$

$$= \pi \left[-\frac{x^5}{20} + \frac{46}{27}x^3 - \frac{7}{3}x^2 + 24x \right]_{-2}^{3} \ldots \approx \boxed{484.474}$$

Example 2

Sketch the region R bounded by the graphs of $y = x^2, y = 4$, and find the volume of the solid generated if R is revolved about the line $y = 4$.

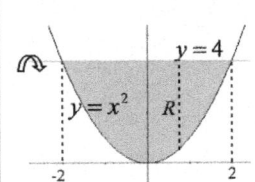

Use Disk Method — since there is no gap between the region and the axis of revolution
$$V = \pi \int_a^b R^2 dx$$

Thickness of disk : dx — since the region is revolved around the line parallel to x-axis

Radius $R = 4 - x^2$

$$V = 2 \cdot \pi \int_0^2 (4 - x^2)^2 dx = 2\pi \int_0^2 (16 - 8x^2 + x^4) dx$$

$$= 2\pi \left[16x - \frac{8}{3}x^3 + \frac{1}{5}x^5 \right]_0^2 = \boxed{\frac{512\pi}{15}}$$

Example 3

Sketch the region R bounded by the graphs of $y = x^2, y = 4$, and find the volume of the solid generated if R is revolved about the line $x = 2$.

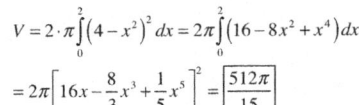

This is the same region as in Example 2, but it is revolved around the different line. So we will get the different solid of revolution.

Use Washer Method:
$$V_{solid} = \pi \int_c^d \left(R_{outer}^2 - r_{inner}^2 \right) dy$$

since there is a gap between the region and the axis of revolution

Thickness of washer : dy — since the region is revolved around the line parallel to y-axis

Example 3 Continued

Outer Radius : $R = 2 - (-\sqrt{y}) = 2 + \sqrt{y}$

Inner Radius : $r = 2 - \sqrt{y}$

$$V = \pi \int_0^4 \left[(2 + \sqrt{y})^2 - (2 - \sqrt{y})^2 \right] dy$$

$$= \pi \int_0^4 8\sqrt{y} \, dy = 8\pi \left[\frac{2}{3} y^{3/2} \right]_0^4 = \boxed{\frac{128\pi}{3}}$$

5.5 Volumes by Cylindrical Shells

Note: Not on the AP Calculus AB or BC Exam

 Mathboat.com

Volumes by Shells

For some types of a solids, it is convenient to use hollow circular cylinders, or thin cylindrical shells to calculate the volume of the solid of revolution.
For Volume of a Solid, add up all the Volumes of each Shell.

The region bounded by $y = f(x)$ and x-axis is rotated around y-axis.

Set up an integral for the volume of the solid of revolution.

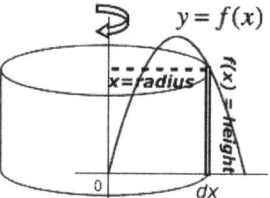

Let's unfold the shell into this flat shape that has a thickness:

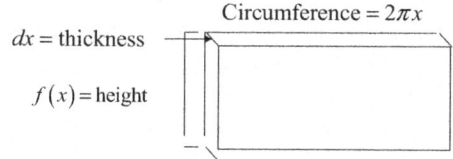

$$V_{\text{one box}} = \text{length} \cdot \text{width} \cdot \text{height} = 2\pi x \cdot f(x) \cdot dx$$

Volume of the solid of revolution is: $\boxed{\text{Volume} = \int_a^b 2\pi x f(x)\, dx}$

$$\boxed{\text{Volume} = \int_a^b 2\pi (\text{radius})(\text{height})\, dx}$$

Tips on Visual Approach

How to draw one shell:

1. Take a point on the curve

2. Draw the radius perpendicular to the axis of rotation

3. Drop a height from the point parallel to the axis of rotation

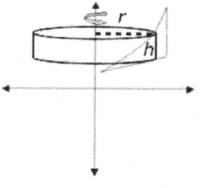

Example 1

The region bounded by the graph of $y = 3x - x^2$ and the x-axis is revolved about the y-axis. Set up the integral to find the volume of the resulting solid.

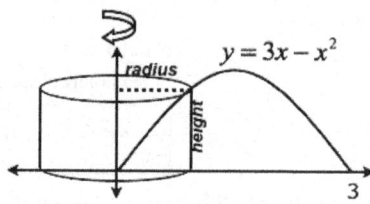

Example 1 Solution

$$\boxed{V = \int_a^b 2\pi (\text{radius})(\text{height})\, dx}$$

Since we build shells along the x-axis, the thickness of Shell: dx

Radius = x

Altitude = $3x - x^2$

$$\boxed{\text{Volume: } \int_0^3 2\pi x (3x - x^2)\, dx}$$

Example 2

The region bounded by the graphs of $y = \frac{1}{2}x^2 + 1$ and $y = x + 5$ is revolved about the line $x = 5$. Set up the integral for the volume of the resulting solid.
Since we build shells along the x-axis, the thickness is dx

Solving for the limits of integration:
$$\frac{1}{2}x^2 + 1 = x + 5$$
$$x^2 - 2x - 8 = 0$$
$$(x-4)(x+2) = 0$$
$$x = 4 \text{ or } -2$$

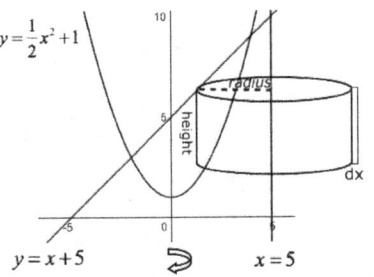

Example 2 Continued

Altitude $= (x+5) - \left(\frac{1}{2}x^2 + 1\right) = -\frac{1}{2}x^2 + x + 4$

Radius $= 5 - x$

Formula: $V = \int_a^b 2\pi (\text{radius})(\text{height}) dx$

$$V = \int_{-2}^{4} 2\pi (5-x)\left(-\frac{1}{2}x^2 + x + 4\right) dx$$

Example 3

The region in the first quadrant bounded by the graph of the equation $x = 3y^3 - y^4$ and the y-axis is revolved about the x-axis. Set up the integral for the volume of the solid.

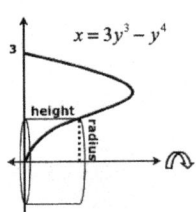

Since we build shells along the y-axis,
Thickness of Shell: dy
Radius $= y$
Altitude $= 3y^3 - y^4$

$$V = \int_a^b 2\pi (\text{radius})(\text{height}) dy$$

$$V = \int_0^3 2\pi y (3y^3 - y^4) dy$$

Example 4

The region below is bounded by the graph $y = \ln x$ and the line $y = x - 2$. Set up an integral expression for the volume of the solid generated when this region is rotated about the y-axis.

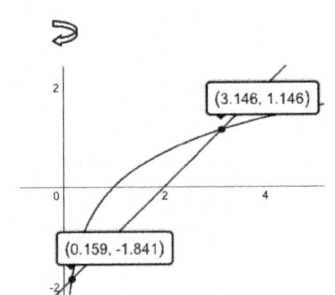

Find points of intersection:
$$e^y = y + 2$$
$$y \approx -1.84141 \text{ and } 1.14619$$

Sometimes, both *Washer* and *Shell* Methods will both work

Example 4 Continued
Using Washer Method:

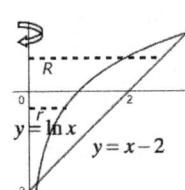

$$V = \pi \int_c^d (R^2 - r^2) dy$$

$R = x_{line} = y + 2$
$r = x_{curve} = e^y$

c and d are the x-coordinates of the points of intersection

$$V = \pi \int_{-1.841}^{1.146} \left((y+2)^2 - (e^y)^2\right) dy$$

Example 4 Continued
Using Shell Method:

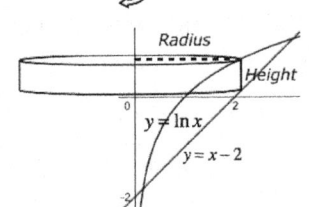

$$V = 2\pi \int_a^b (\text{radius})(\text{height}) dx$$

Radius $= x$
Height $= (\ln x) - (x - 2)$

a and b are the x-coordinates of the points of intersection

$$V = 2\pi \int_{0.159}^{3.146} x((\ln x) - (x-2)) dx$$

If you use graphing calculator on both integral expressions for the volume, answers will be the same.

5.6 Volumes by Cross Sections

 Mathboat.com

Volumes by Cross Sections

Let *S* be a solid bounded by planes that are perpendicular to the *x*-axis at *a* and *b*.

If, for every *x* in [*a*, *b*], the cross-sectional area of *S* is given by *A*(*x*), where *A* is continuous on [*a*,*b*], then the volume of *S* is:

$$V = \int_a^b A(x)\,dx$$

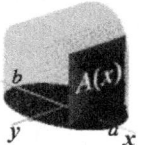

Example 1

Find the volume of a right pyramid with a square base of side *a* and and altitude *h*.

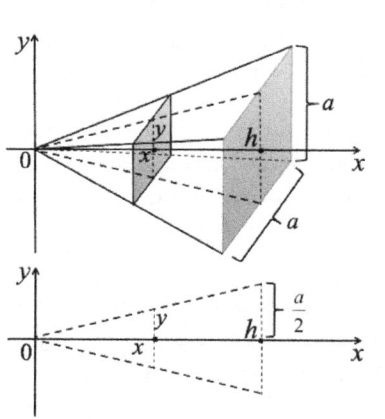

Example 1 Continued

$$V = \int_a^b A(x)\,dx$$

Cross sectional area equals

$$A(x) = (2y)^2 = 4y^2$$

$$\frac{y}{x} = \frac{\frac{a}{2}}{h} \quad \text{use similar triangles}$$

$$y = \frac{ax}{2h} \quad \text{solve for } y$$

Example 1 Continued

$$A(x) = 4y^2 = \frac{4a^2 x^2}{4h^2} = \frac{a^2}{h^2} x^2 \quad \text{plug in } y \text{ and simplify}$$

$$V = \int_0^h A(x)\,dx = \int_0^h \boxed{\frac{a^2}{h^2}} x^2 \,dx = \quad \begin{array}{l}\text{use the formula of}\\ \text{Volume by Cross}\\ \text{Sections}\end{array}$$

$$\boxed{\frac{a^2}{h^2}} \cdot \left[\frac{x^3}{3}\right]_0^h = \frac{a^2 h^3}{h^2 \cdot 3} = \boxed{\frac{1}{3} a^2 h} \quad \begin{array}{l}\text{when you integrate,}\\ \text{take the constant out}\end{array}$$

Example 2

Find the volume of the solid whose base is bounded by the graphs of $y = x+1$ and $y = x^2-1$, with **square** cross sections taken perpendicular to the **x-axis**.

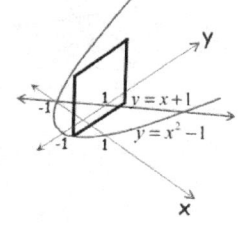

Guidelines for sketching

1. Draw set of axis. Put letters *x* and *y* on (+) sides of *x*-axis and *y*-axis
2. Draw line $y = x+1$
3. Draw curve $y = x^2-1$
4. Since cross-section is perpendicular to *x*-axis, draw the line through the region, which is parallel to *y*-axis
5. Draw vertical heights of your cross- section
6. Connect them with the line segment parallel to the line segment from # 4

Example 2 Continued

$$V = \int_a^b A(x)\,dx$$

Square's side length =
y of upper curve – y of lower curve
$x+1-(x^2-1) = -x^2+x+2$

Square's area $= (-x^2+x+2)^2$

limits for integration = intersection points
$x+1 = x^2-1$
$x^2-x-2 = 0$
$(x+1)(x-2) = 0,\ x = -1, 2$

$\int_{-1}^{2}(-x^2+x+2)^2\,dx = \int_{-1}^{2}(x^4 - 2x^3 - 3x^2 + 4x + 4)\,dx$

note: $(a+b+c)^2 = a^2+b^2+c^2 + 2ab+2bc+2ac$

$= \left[\dfrac{x^5}{5} - \dfrac{x^4}{2} - x^3 + 2x^2 + 4x\right]_{-1}^{2} = \dfrac{81}{10} = \boxed{8.1}$

Example 3

Example of drawing. Draw the solid whose base is bounded by the graphs of $y = x+1$ and $y = x^2-1$, with **Equilateral Triangle** cross sections taken perpendicular to the *x*-axis.

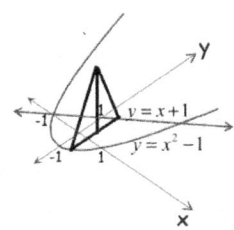

1. Draw set of axis. Put letters x and y on (+) sides of x-axis and y-axis
2. Draw line $y = x+1$
3. Draw curve $y = x^2-1$
4. Since cross-section is perpendicular to x-axis, draw the segment through the region, which is parallel to y-axis
5. At the midpoint of this segment draw vertical height of your cross-section
6. Construct the triangle

Example 4

Example of drawing. Draw the solid whose base is bounded by the graphs of $y = x+1$ and $y = x^2-1$, with **rectangular** cross sections taken perpendicular to the **y-axis**.

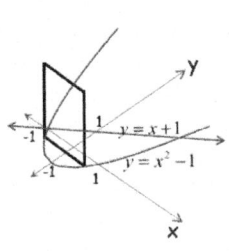

1. Draw set of axis. Put letters x and y on (+) sides of x-axis and y-axis
2. Draw line $y = x+1$
3. Draw curve $y = x^2-1$
4. Since cross-section is perpendicular to y-axis, draw the segment through the region, which is parallel to x-axis
5. Draw vertical heights of your cross-section
6. Connect them with the line segment parallel to the line segment from # 4

Examples on Sketching Cross Sections

A solid has as its base the circular region in the xy-plane bounded by the graph of $x^2 + y^2 = a^2$.

Draw the diagram if:

Every cross section by a plane $\perp$ to the *x*-axis is a square.

Every cross section by a plane $\perp$ to the *x*-axis is an isoseles right triangle with hypotenuse on the *xy*-plane.

Example on Sketching a Cross Section

Every cross section by a plane $\perp$ to the *x*-axis is a semicircle with diameter in *xy* plane.

Example 5

Find the volume of the solid whose base is the region bounded by the semicircle $y = \sqrt{9-x^2}$ and the *x*-axis and whose cross-sections by a plane perpendicular to the *x*-axis are equilateral triangles.

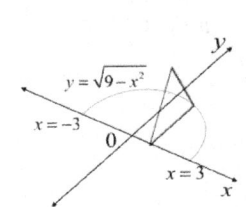

Area of Equilateral Triangle $= a^2 \dfrac{\sqrt{3}}{4}$
where *a* is the length of a side of a triangle.
$a = \sqrt{9-x^2}$

Area of one cross section: $A(x) = \dfrac{\sqrt{3}}{4}(9-x^2)$

Limits of integration: $x = -3,\ x = 3$
Integrate with respect to *x*:

$$V = \int_a^b A(x)\,dx = \boxed{\int_{-3}^{3} \dfrac{\sqrt{3}}{4}(9-x^2)\,dx}$$

5.7 Arc Length
Note: Not on the AP Calculus AB Exam

Arc Length Formula

Let f be smooth(differentiable) on $[a,b]$.

The arc length of the graph of f from $A(a, f(a))$ to $B(b,f(b))$ is

$$L_a^b = \int_a^b \sqrt{1+[f'(x)]^2}\,dx$$

Arc Length Formula

$$L_a^b = \int_a^b \sqrt{1+[f'(x)]^2}\,dx$$

$\dfrac{\Delta y_k}{\Delta x_k} = \dfrac{\text{change in } y}{\text{change in } x} = \text{slope} = \dfrac{dy}{dx}$

Let's pick some points on the arc AB and connect them with line segments. If we pick a lot of points, the sum of the lengths of these segments will approximately be equal to the length of arc AB.

Proof of Arc Length formula

Length of arc AB: L_a^b = limit of Riemann sums when the largest Δx_k, which is called the norm of the partition, $\|p\| \to 0$

$$L_a^b = \lim_{\|p\|\to 0} \sum \sqrt{(\Delta x_k)^2 + (\Delta y_k)^2} =$$

$$\lim_{\|p\|\to 0} \sum \sqrt{\frac{(\Delta x_k)^2 + (\Delta y_k)^2}{(\Delta x_k)^2} \cdot (\Delta x_k)^2} =$$

$$\lim_{\|p\|\to 0} \sum \sqrt{\frac{(\Delta x_k)^2}{(\Delta x_k)^2} + \frac{(\Delta y_k)^2}{(\Delta x_k)^2}} \cdot (\Delta x_k) =$$

$$\lim_{\|p\|\to 0} \sum \sqrt{1 + \frac{(\Delta y_k)^2}{(\Delta x_k)^2}} \cdot (\Delta x_k) = \int_a^b \sqrt{1+\left(\frac{dy}{dx}\right)^2}\,dx = \int_a^b \sqrt{1+(f'(x))^2}\,dx$$

Example 1

If $f(x) = 3x^{2/3} + 4$, find the arc length of the graph of f from the point $A(8,16)$ to $B(27, 31)$.

$$L_a^b = \int_a^b \sqrt{1+[f'(x)]^2}\,dx \quad \text{You don't need to graph it!}$$

$$f'(x) = 3 \cdot \frac{2}{3} x^{-\frac{1}{3}} = 2x^{-\frac{1}{3}}$$

$$L_8^{27} = \int_8^{27} \sqrt{1+\left(\frac{2}{x^{1/3}}\right)^2}\,dx = \int_8^{27} \sqrt{1+\frac{4}{x^{2/3}}}\,dx$$

Example 1 Continued

$$L_8^{27} = \int_8^{27} \sqrt{\frac{x^{2/3}+4}{x^{2/3}}}\,dx$$

$$= \int_8^{27} \left(\sqrt{x^{2/3}+4} \cdot \frac{1}{x^{1/3}}\right)dx$$

$$= \frac{3}{2}\int_8^{13} \sqrt{u}\,du = u^{3/2}\Big]_8^{13}$$

$$= 13^{3/2} - 8^{3/2} \approx \boxed{24.245}$$

$u = x^{2/3} + 4$
$du = \frac{2}{3}x^{-1/3}dx$
$\frac{1}{x^{1/3}}dx = \frac{3}{2}du$

$x = 8 \to u = 8^{2/3} + 4 = 8$
$x = 27 \to u = 27^{2/3} + 4 = 13$

Arc Length Formula (in terms of y)

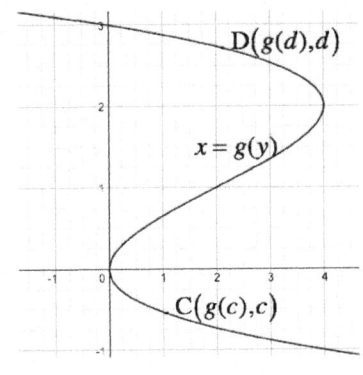

$D(g(d),d)$
$x = g(y)$
$C(g(c),c)$

$$L_c^d = \int_c^d \sqrt{1+[g'(y)]^2}\,dy$$

The proof is similar to the one in terms of x

Example 2

Set up an integral for finding the arc length of the graph of the equation $2y^3 + 4y - x = 0$ from $A(0,0)$ to $B(24,2)$.

Since in the given equation it is easier to express x in terms of y, we'll use this formula of the length of arc:

$$L_c^d = \int_c^d \sqrt{1+[g'(y)]^2}\,dy$$

$x = 2y^3 + 4y;\ c=0,\ d=2$

$g'(y) = (2y^3+4y)' = 6y^2+4$

$$L_0^2 = \int_0^2 \sqrt{1+(6y^2+4)^2}\,dy$$

$$= \int_0^2 \sqrt{36y^4+48y^2+17}\,dy$$

Example 3

Find the length of the curve defined by the equation

$f(x) = \dfrac{x^3}{12} + \dfrac{1}{x}$ from $A\left(1, \dfrac{13}{12}\right)$ to $B\left(2, \dfrac{7}{6}\right)$

$$L_a^b = \int_a^b \sqrt{1+[f'(x)]^2}\,dx$$

$f'(x) = \dfrac{1}{12}\cdot 3x^2 - \dfrac{1}{x^2} = \dfrac{1}{4}x^2 - \dfrac{1}{x^2}$

$$L_1^2 = \int_1^2 \sqrt{1+\left(\dfrac{1}{4}x^2 \boxed{-} \dfrac{1}{x^2}\right)^2}\,dx$$

$$= \int_1^2 \sqrt{1+\dfrac{1}{16}x^4 \boxed{-} \dfrac{1}{2} + \dfrac{1}{x^4}}\,dx$$

$$= \int_1^2 \sqrt{\dfrac{1}{16}x^4 \boxed{+} \dfrac{1}{2} + \dfrac{1}{x^4}}\,dx$$

Example 3 Continued

$$L_1^2 = \int_1^2 \sqrt{\left(\dfrac{1}{4}x^2 \boxed{+} \dfrac{1}{x^2}\right)^2}\,dx = \int_1^2 \left(\dfrac{1}{4}x^2 + \dfrac{1}{x^2}\right)dx$$

$$= \dfrac{1}{4}\dfrac{x^3}{3} + \dfrac{x^{-1}}{-1}\Big]_1^2 = \dfrac{1}{12}x^3\Big]_1^2 - \dfrac{1}{x}\Big]_1^2$$

$$= \dfrac{1}{12}(8-1) - \left(\dfrac{1}{2}-1\right) = \dfrac{7}{12} + \dfrac{1}{2} = \boxed{\dfrac{13}{12}}$$

Use this space for notes.

Use this space for notes.

6.1 INVERSE FUNCTIONS

Definition

A function f with domain D and range R is a one-to-one function if whenever $a \ne b$ in D, then $f(a) \ne f(b)$ in R

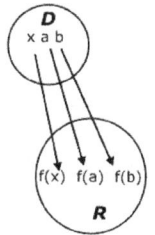

Definition

If $f(x)$ is one-to-one function with domain D and range R, then a function g with domain R and range D is its inverse function with the condition:
$y = f(x) \Leftrightarrow x = g(y)$ for every x in D and y in R

Horizontal Line Test

If the graph of a function y = f(x) is such that no horizontal line intersects the graph in more than one point, then it is one to one and has an inverse function.

One-to-one function, has an Inverse Function

Not One-to-one function, does not have an Inverse Function

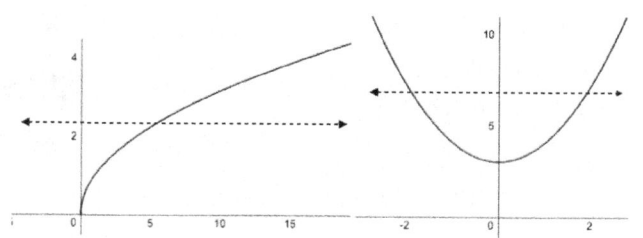

Definitions

$f(x)$ and $g(x)$ are Inverse Functions if:
1) $g(f(x)) = x$ for all x in the domain of f
2) $f(g(x)) = x$ for all x in the domain of g

Example 1

Given: $f(x) = \dfrac{x-1}{2}$ and $g(x) = 2x+1$.

a) Show that f and g are inverses of each other.

$g(f(x)) = g\left(\dfrac{x-1}{2}\right) = 2 \cdot \left(\dfrac{x-1}{2}\right) + 1 = x$ for all x

$f(g(x)) = f(2x+1) = \dfrac{(2x+1)-1}{2} = x$ for all x

$\therefore g = f^{-1}$, inverse of f

Example 1 Continued

b) Graph f and g.

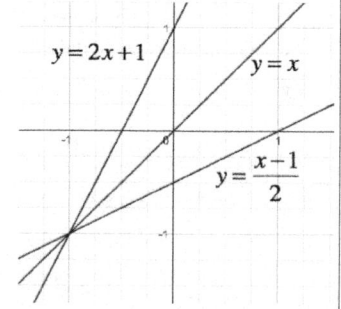

For every point (a,b) on the graph of f the point (b,a) is on the the graph of f⁻¹.

The graph of f⁻¹ is the reflection of graph of f in line y = x

Example 2

Given: $f(x) = 2x - 5$.

Find and graph $f^{-1}(x)$.

$y = 2x - 5$

Replace y with x

$x = 2y - 5$

Solve for y

$x + 5 = 2y$

$\boxed{y = \dfrac{x+5}{2}}$

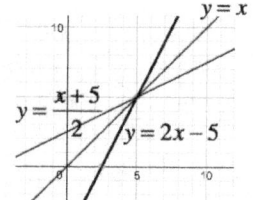

Example 3

Given: $f(x) = 9 - x^2$ for $x \geq 0$.

Find and graph $f^{-1}(x)$.

$y = 9 - x^2$

Replace y with x

$x = 9 - y^2$; $\boxed{y \geq 0}$

Solve for y

$y^2 = 9 - x$

$\boxed{y = \sqrt{9-x}}$

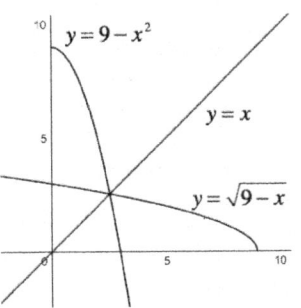

Guidelines for finding f^{-1}

1. Verify that f is a one-to-one function (or that f is increasing or decreasing) on its domain.
2. Solve the equation $y = f(x)$ for x in terms of y, obtaining $x = f^{-1}(y)$
3. Verify the two conditions $f^{-1}(f(x)) = x$ and $f(f^{-1}(x)) = x$ for every x in the domains of f and f^{-1} respectively.

Theorems

If f is continuous and increasing on $[a,b]$, then f has an inverse function f^{-1} that is continuous and increasing on $[f(a), f(b)]$.

If a differentiable function f has an inverse $g = f^{-1}$ and if $f'(g(c)) \neq 0$, then g is differentiable at c and $g'(c) = \dfrac{1}{f'(g(c))}$

$\boxed{g'(c) = \dfrac{1}{f'(g(c))}}$ proof

f and g are inverses of each other $\Rightarrow$

$(f(g(x))) = x$

Take derivative on both sides:

$(f(g(x)))' = x'$

By chain rule: $\downarrow$

$f'(g(x)) \cdot g'(x) = 1$

$g'(x) = \dfrac{1}{f'(g(x))} \Rightarrow \boxed{g'(c) = \dfrac{1}{f'(g(c))}}$

Example 4

If $f(x) = x^3 + 2x$, prove that f has an inverse function g, and find the slope of the tangent line to the graph of g at the point $\boxed{P(3,1)}$

So, $\boxed{g(3) = 1, g'(3) = ?}$

$f'(x) = 3x^2 + 2 > 0$ for every $x \Rightarrow f$ is increasing

$\Rightarrow f$ is one-to-one $\Rightarrow f$ has an inverse function g

$\boxed{g'(c) = \dfrac{1}{f'(g(c))}} \Rightarrow g'(3) = \dfrac{1}{f'(g(3))}$

$g(3) = 1 \Rightarrow g'(3) = \dfrac{1}{f'(g(3))} = \dfrac{1}{f'(1)} = \boxed{\dfrac{1}{5}}$

Example 5 (when g(3) is not given)

If $f(x) = x^3 + 2x$, prove that f has an inverse function g, and find the slope of the tangent line to the graph of g at $\boxed{x=3}$. So, $\boxed{g(3) \text{ is uknown}, g'(3) = ?}$

For $g: x = 3, y = ? \Rightarrow g(3) = ?$

For $f: y = 3, x = ? \Rightarrow f(?) = 3 \Rightarrow$ Set $f(x) = 3$

$x^3 + 2x = 3 \Rightarrow x^3 + 2x - 3 = 0 \Rightarrow x = 1 \Rightarrow \boxed{g(3) = 1}$

$f'(x) = 3x^2 + 2 > 0$ for every $x \Rightarrow f$ is increasing

$\Rightarrow f$ is one-to-one $\Rightarrow f$ has an inverse function g

$\boxed{g'(c) = \dfrac{1}{f'(g(c))}} \Rightarrow g'(3) = \dfrac{1}{f'(g(3))}$

$g(3) = 1 \Rightarrow g'(3) = \dfrac{1}{f'(g(3))} = \dfrac{1}{f'(1)} = \boxed{\dfrac{1}{5}}$

6.2 THE NATURAL LOGARITHMIC FUNCTION

Mathboat.com

Definition

The natural logarithmic function denoted by ln, is defined by $\ln x = \int_1^x \frac{1}{t} dt$ for every $x > 0$

Let's investigate function $y = \ln x$ and then graph it.

If $x > 1$, $\int_1^x \frac{1}{t} dt = \ln x$ is the area of the region under $y = \frac{1}{t}$ from $t = 1$ to $t = x$

If $0 < x < 1$, $\int_1^x \frac{1}{t} dt = -\int_x^1 \frac{1}{t} dt$
= negative of the area of region under $y = \frac{1}{t}$ from $t = x$ to $t = 1$

$\ln x < 0$ for $0 < x < 1$
$\ln x > 0$ for $x > 1$

Graph of y = ln x

By 1st Fundamental Theorem of Calculus,

$\frac{d}{dx}\int_1^x \frac{1}{t} dt = \frac{1}{x}$

So, $\frac{d}{dx} \ln x = \frac{1}{x}$

$(\ln x)' = \frac{1}{x}$

$\ln x < 0$ for $0 < x < 1$
$\ln x > 0$ for $x > 1$

$y = \ln x$

$(\ln x)' = \frac{1}{x} > 0 \Rightarrow \ln x$ is increasing function
$(\ln x)'' = -\frac{1}{x^2} < 0 \Rightarrow \ln x$ is concave downward

Theorem

$(\ln|u|)' = \frac{1}{u}, \; u \neq 0$

Proof

$u > 0 \Rightarrow (\ln u)' = \frac{1}{u}$

$u < 0 \Rightarrow$ By Chain Rule:

$(\ln(-u))' = \frac{1}{-u} \cdot -1 = \frac{1}{u}$

$(\ln|u|)' = \frac{1}{u}$

LAWS OF LOGARITHMS

If $p > 0$ and $q > 0$, then

$\ln pq = \ln p + \ln q$

$\ln \frac{p}{q} = \ln p - \ln q$

$\ln p^r = r \ln p$

Example 1

Find the derivative of the function:

$y = \ln(x^2 + 2x + 1)^{\frac{1}{3}}$

Simplify first

$y = \frac{1}{3}\ln(x+1)^2$

Put exponent 2 in front of ln

$y = \frac{2}{3}\ln(x+1)$

$y' = \frac{2}{3} \cdot \frac{1}{x+1} \cdot 1 = \boxed{\frac{2}{3(x+1)}}$

Example 2

Find the derivative of the function:

$$y = \ln \sqrt[3]{\frac{x^2-4}{x^2+4}}$$

Simplify first: $\quad y = \ln\left(\frac{x^2-4}{x^2+4}\right)^{\frac{1}{3}}$

Put $\frac{1}{3}$ in front of ln: $\quad y = \frac{1}{3}\ln\left(\frac{x^2-4}{x^2+4}\right)$

Now use: $\ln\frac{p}{q} = \ln p - \ln q$:

$$y = \frac{1}{3}\left(\ln(x^2-4) - \ln(x^2+4)\right)$$

Example 2 Continued

By Chain rule, $\quad \dfrac{dy}{dx} = \dfrac{1}{3}\left(\dfrac{1}{x^2-4}\cdot 2x - \dfrac{1}{x^2+4}\cdot 2x\right)$

$$= \frac{2x}{3}\left(\frac{1}{x^2-4} - \frac{1}{x^2+4}\right)$$

$$= \frac{2x}{3}\left(\frac{x^2+4-x^2+4}{(x^2+4)(x^2-4)}\right)$$

$$= \frac{2x}{3}\left(\frac{8}{(x^2+4)(x^2-4)}\right)$$

$$\boxed{\frac{dy}{dx} = \frac{16x}{3(x^2+4)(x^2-4)}}$$

Example 3

Find the derivative of the function defined by equation:

$3y - x^2 + \ln xy = 2$

Use: $\ln(pq) = \ln p + \ln q$

$3y - x^2 + \ln x + \ln y = 2$

Implicitly differentiate

$3\dfrac{dy}{dx} - 2x + \dfrac{1}{x} + \dfrac{1}{y}\cdot\dfrac{dy}{dx} = 0$

Factor out $\dfrac{dy}{dx}$

$\dfrac{dy}{dx}\left(3 + \dfrac{1}{y}\right) = 2x - \dfrac{1}{x}$

$\dfrac{dy}{dx} = \dfrac{\dfrac{2x^2-1}{x}}{\dfrac{3y+1}{y}}$

$\boxed{\dfrac{dy}{dx} = \dfrac{(2x^2-1)y}{(3y+1)x}}$

GUIDELINES FOR LOGARITHMIC DIFFERENTIATION

$y = f(x)$

$\ln y = \ln f(x)$ Take ln on both sides

$\dfrac{1}{y}\cdot\dfrac{dy}{dx} = (\ln f(x))'$ Implicitly differentiate

$\dfrac{dy}{dx} = y\cdot(\ln f(x))'$ Multiply both sides by y

$\dfrac{dy}{dx} = f(x)\cdot(\ln f(x))'$ Replace y with f(x)

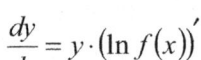

Example 4

Find derivative of the function:

$y = \dfrac{(3x-2)^4}{\sqrt{4x+1}} \Rightarrow$ Take ln on both sides

$\ln y = \ln\dfrac{(3x-2)^4}{\sqrt{4x+1}} \Rightarrow$ Simplify

$\ln y = \ln(3x-2)^4 - \ln\sqrt{4x+1}$

$\ln y = 4\ln(3x-2) - \dfrac{1}{2}\ln(4x+1)$

Implicitly differentiate

$(\ln y)' = (4\ln(3x-2))' - \left(\dfrac{1}{2}\ln(4x+1)\right)'$

$\dfrac{1}{y}\dfrac{dy}{dx} = 4\cdot\dfrac{1\cdot 3}{3x-2} - \dfrac{1}{2}\cdot\dfrac{1\cdot 4}{4x+1} = \dfrac{42x+16}{(3x-2)(4x+1)}$

Example 4 Continued

Multiply both sides by y

$\dfrac{dy}{dx} = y\cdot\dfrac{42x+16}{(3x-2)(4x+1)}$

Replace y with original function

$\dfrac{dy}{dx} = \dfrac{(3x-2)^4}{\sqrt{4x+1}}\cdot\dfrac{42x+16}{(3x-2)(4x+1)}$

$\dfrac{dy}{dx} = \dfrac{(3x-2)^3(42x+16)}{(4x+1)^{\frac{3}{2}}}$

6.3 THE NATURAL EXPONENTIAL FUNCTION

 Mathboat.com

The Natural Exponential Function

The natural exponential function $y = e^x$ is the inverse of the natural logarithmic function $y = \ln x$

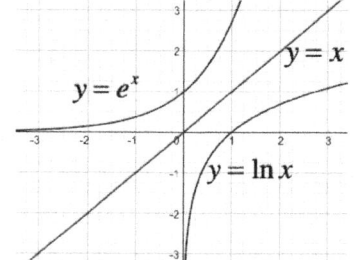

Definition.

If x is any real number,

$$e^x = y \Leftrightarrow \ln y = x$$

$$e \approx 2.71828$$

$\Rightarrow$ irrational number

Later, we will show that $e = \lim_{h \to 0}(1+h)^{\frac{1}{h}}$

Theorems

$\boxed{\ln e^x = x \text{ for every } x}$ We proved them in Pre-Calculus $\boxed{e^{\ln x} = x \text{ for every } x > 0}$

Examples:

$\ln e^{\sqrt{x+2}} = \sqrt{x+2}$

$e^{\ln\sqrt{3x+1}} = \sqrt{3x+1}$

$e^{5\ln x} = \left(e^{\ln x}\right)^5 = x^5$

Theorems

If p and q are real numbers and r is a rational number, then

$$e^p e^q = e^{p+q}$$

$$\frac{e^p}{e^q} = e^{p-q}$$

$$\left(e^p\right)^q = e^{pq}$$

Theorem

$$\boxed{\left(e^x\right)' = e^x}$$

proof:

$\ln e^x = x$

$\left(\ln e^x\right)' = x'$ take derivative on both sides

$\frac{1}{e^x}\left(e^x\right)' = 1$ don't forget the Chain Rule

$\left(e^x\right)' = e^x$

Example 1

If $f(x) = x^2 e^x$, find $f'(x)$

Solution

Use Product Rule

$f'(x) = x^2\left(e^x\right)' + e^x\left(x^2\right)'$

$= x^2 e^x + e^x(2x)$

$\boxed{= xe^x(x+2)}$

Example 2

Find $\left(e^{\sqrt{3x^2+2x+4}}\right)'$.

Use Chain Rule

$$\left(e^{\sqrt{3x^2+2x+4}}\right)' = e^{\sqrt{3x^2+2x+4}}\left(\sqrt{3x^2+2x+4}\right)' =$$

$$= e^{\sqrt{3x^2+2x+4}} \cdot \frac{(6x+2)}{2\sqrt{3x^2+2x+4}}$$

$$= \frac{(3x+1)e^{\sqrt{3x^2+2x+4}}}{\sqrt{3x^2+2x+4}}$$

Example 3

If $e^{2x-3xy} = 4$, then $\dfrac{dy}{dx} = ?$

Implicitly differentiate:

$$e^{2x-3xy} \cdot \big(2 - 3(x'y + y'x)\big) = 0$$

$$e^{2x-3xy}\left(2 - 3y - 3x\frac{dy}{dx}\right) = 0$$

Example 3 Continued

$e^{2x-3xy} \neq 0$, always positive

$$2 - 3y - 3x\frac{dy}{dx} = 0$$

$$\frac{dy}{dx} = \boxed{\frac{2-3y}{3x}}$$

Example 4

Given $f(x) = e^{-\frac{x^2}{2}}$. Find points of local extrema, concavity and points of inflection.

$$f'(x) = e^{-\frac{x^2}{2}}\left(-\frac{x^2}{2}\right)' = e^{-\frac{x^2}{2}}\left(-\frac{2x}{2}\right) = -xe^{-\frac{x^2}{2}}$$

$$f'(x) = 0 \Rightarrow -xe^{-\frac{x^2}{2}} = 0$$

$$e^{-\frac{x^2}{2}} > 0, x = 0$$

$\Rightarrow x = 0$ is the critical number.

$f'(x)$: $+$ Local max $-$ at 0

point of local max $= (0, f(0)) = (0, e^0) = \boxed{(0,1)}$

Example 4 Continued

$$f''(x) = \big(f'(x)\big)' = \left(-xe^{-\frac{x^2}{2}}\right)'$$

$$= -\left(x'e^{-\frac{x^2}{2}} + x\cdot\left(e^{-\frac{x^2}{2}}\right)'\right)$$

$$= -\left(e^{-\frac{x^2}{2}} - x\cdot xe^{-\frac{x^2}{2}}\right)$$

$$= e^{-\frac{x^2}{2}}(x^2-1) = 0$$

$$e^{-\frac{x^2}{2}} > 0, x^2 - 1 = 0$$

$$x = 1,\ x = -1$$

Example 4 Continued

$f''(x)$: $+$ $\smiley$ -1 $-$ $\frownie$ 1 $+$ $\smiley$

$\boxed{\text{CU at } (-\infty,-1)\cup(1,+\infty),\ \text{CD at } (-1,1)}$

Points of Inflections:

$$\left(-1, e^{-\frac{(-1)^2}{2}}\right) = \boxed{(-1, 0.6065)}$$

$$\left(1, e^{-\frac{(1)^2}{2}}\right) = \boxed{(1, 0.6065)}$$

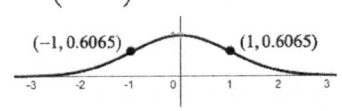

6.4 Integration.
Integrals of $1/x$ and some trig functions

Theorem
If $u = g(x) \neq 0$ and g is differentiable, then
$$\int \frac{1}{u} du = \ln|u| + C$$

Proof:

$$(\ln|u|)' = \frac{1}{u} \qquad \text{We proved it before}$$

$$\int \frac{1}{u} du = \ln|u| + C \qquad \text{Use differentiation formula for ln to obtain formula for integration}$$

Example 1

Evaluate: $\int \frac{x}{5x^2 + 6} dx$

$\boxed{\begin{array}{l} u = 5x^2 + 6 \\ du = 10x\, dx \\ x\, dx = \dfrac{du}{10} \end{array}}$

$\dfrac{1}{10} \int \dfrac{1}{u} du =$

$\dfrac{1}{10} \ln|u| + C = \boxed{\dfrac{1}{10} \ln\left|5x^2 + 6\right| + C}$

Example 2

Evaluate: $\int_0^4 \dfrac{1}{10 - 2x} dx$

$\boxed{\begin{array}{l} u = 10 - 2x \\ du = -2\, dx \\ dx = -\dfrac{du}{2} \end{array}}$

$= -\dfrac{1}{2} \int_{10}^{2} \dfrac{1}{u} du = -\dfrac{1}{2} \Big[\ln|u|\Big]_{10}^{2}$

$\boxed{\begin{array}{l} x = 0 \Rightarrow u = 10 - 2 \cdot 0 = 10 \\ x = 4 \Rightarrow u = 10 - 2 \cdot 4 = 2 \end{array}}$

$= -\dfrac{1}{2} (\ln 2 - \ln 10)$

$= -\dfrac{1}{2} \ln \dfrac{2}{10} = \dfrac{1}{2} \ln 5$

$= -\dfrac{1}{2} \ln \dfrac{2}{10} = \boxed{\dfrac{1}{2} \ln 5}$

Example 3

$\int \dfrac{t^2}{3t + 3} dt =$

Add and subtract 1 in numerator and factor out 3 in denominator

$\dfrac{1}{3} \int \dfrac{t^2 - 1 + 1}{t + 1} dt = \dfrac{1}{3} \int \dfrac{t^2 - 1}{t + 1} dt + \dfrac{1}{3} \int \dfrac{1}{t + 1} dt =$

$\dfrac{1}{3} \int \dfrac{(t - 1)\cancel{(t + 1)}}{\cancel{t + 1}} dt + \dfrac{1}{3} \int \dfrac{1}{t + 1} dt =$

Example 3 Continued

$\dfrac{1}{3} \int (t - 1) dt + \dfrac{1}{3} \int \dfrac{1}{t + 1} dt =$

$\boxed{\dfrac{1}{3} \left(\dfrac{t^2}{2} - t + \ln|t + 1| \right) + C}$

Example 4

Evaluate: $\int \dfrac{\sqrt{\ln(2x)}}{5x}dx$

$u = \ln(2x)$
$du = \dfrac{1}{2x}\cdot 2dx = \dfrac{dx}{x}$

$= \dfrac{1}{5}\int u^{\frac{1}{2}}du$

$= \dfrac{1}{5}\cdot\dfrac{2}{3}u^{\frac{3}{2}}+C = \dfrac{2}{15}(\ln(2x))^{\frac{3}{2}}+C$

Theorem

If $u = g(x)$ and g is differentiable, then
$\int e^u\,du = e^u + C$

Proof

$(e^u)' = e^u$ — We proved it before

$\int e^u\,du = e^u + C$ — Use differentiation formula for e^u to obtain formula for integration

Example 5

a) Evaluate

$\int \dfrac{e^{\frac{4}{x}}}{5x^2}dx$

$u = \dfrac{4}{x}$
$du = -\dfrac{4}{x^2}dx$
$\dfrac{dx}{x^2} = -\dfrac{du}{4}$

$= -\dfrac{1}{20}\int e^u\,du$

$= -\dfrac{1}{20}e^u + C$

$= -\dfrac{1}{20}e^{\frac{4}{x}} + C$

b) Evaluate

$\int_1^3 \dfrac{e^{\frac{4}{x}}}{5x^2}dx$

$= -\dfrac{1}{20}\left[e^{\frac{4}{x}}\right]_1^3$

$= -\dfrac{1}{20}\left(e^{\frac{4}{3}} - e^4\right)$

Example 6

Find y given $\dfrac{dy}{dx} = 6e^{4x} + 2e^{-5x}$ if $y = 5$ when $x = 0$.

$\int dy = \int (6e^{4x} + 2e^{-5x})dx$

$y = \dfrac{6}{4}e^{4x} + \dfrac{2}{-5}e^{-5x} + C$

$5 = \dfrac{3}{2}e^0 - \dfrac{2}{5}e^0 + C = \dfrac{3}{2} - \dfrac{2}{5} + C \;\Rightarrow\; C = \dfrac{39}{10} = 3.9$

$y = \dfrac{3}{2}e^{4x} - \dfrac{2}{5}e^{-5x} + 3.9$

Example 7

Find the area of the region bounded by the graphs of the equations: $y = e^{2x}, y = \sqrt{x+4}, x = -3, x = -1$

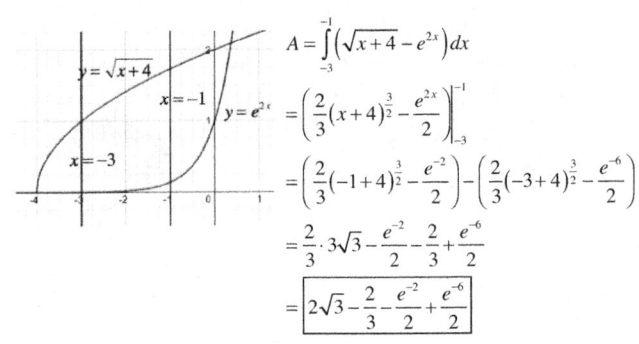

$A = \int_{-3}^{-1}(\sqrt{x+4} - e^{2x})dx$

$= \left(\dfrac{2}{3}(x+4)^{\frac{3}{2}} - \dfrac{e^{2x}}{2}\right)\Big|_{-3}^{-1}$

$= \left(\dfrac{2}{3}(-1+4)^{\frac{3}{2}} - \dfrac{e^{-2}}{2}\right) - \left(\dfrac{2}{3}(-3+4)^{\frac{3}{2}} - \dfrac{e^{-6}}{2}\right)$

$= \dfrac{2}{3}\cdot 3\sqrt{3} - \dfrac{e^{-2}}{2} - \dfrac{2}{3} + \dfrac{e^{-6}}{2}$

$= 2\sqrt{3} - \dfrac{2}{3} - \dfrac{e^{-2}}{2} + \dfrac{e^{-6}}{2}$

Theorem

(1) $\int \tan u\,du = -\ln|\cos u| + C$ or $\int \tan u\,du = \ln|\sec u| + C$

(2) $\int \cot u\,du = \ln|\sin u| + C$ or $\int \cot u\,du = -\ln|\csc u| + C$

(3) $\int \sec u\,du = \ln|\sec u + \tan u| + C$

(4) $\int \csc u\,du = \ln|\csc u - \cot u| + C$

Let's prove (1) and (3)

$$\int \tan x\, dx = -\ln|\cos x| + C$$

Proof

$$\int \tan x\, dx = \int \frac{\sin x}{\cos x}\, dx$$

$u = \cos x$
$du = -\sin x\, dx$
$\cos x \neq 0$

$$= -\int \frac{du}{u} = -\ln|u| + c$$

$$= -\ln|\cos x| + C$$

$$\int \sec x\, dx = \ln|\sec x + \tan x| + C$$

Proof

$u = \sec x + \tan x$
$du = (\sec x \tan x + \sec^2 x)\, dx$

$$\int \sec x\, dx = \int \sec x \cdot \frac{\sec x + \tan x}{\sec x + \tan x}\, dx$$

Multiply and divide by $(\sec x + \tan x)$

$$= \int \frac{\sec^2 x + \sec x \tan x}{\sec x + \tan x}\, dx$$

Simplify numerator: Multiply $\sec x$ by $(\sec x + \tan x)$

$$= \int \frac{du}{u} = \ln|u| + C = \ln|\sec x + \tan x| + C$$

Example 8

Evaluate $\int x \cot x^2\, dx$

$u = x^2$
$du = 2x\, dx$
$x\, dx = \frac{1}{2} du$

$$= \frac{1}{2}\int \cot u\, du$$

$$= \frac{1}{2}\ln|\sin u| + C$$

$$= \frac{1}{2}\ln|\sin x^2| + C$$

Use this space for notes.

Use this space for notes.

Use this space for notes.

6.5 General Exponential and Logarithmic Functions

Definition
If $f(x) = a^x$, then f is the exponential function with base a

Laws of Exponents:

Let $a > 0$ and $b > 0$. If u and v are any real numbers, then

$$a^u a^v = a^{u+v} \qquad (a^u)^v = a^{uv} \qquad (ab)^u = a^u b^u$$

$$\frac{a^u}{a^v} = a^{u-v} \qquad \left(\frac{a}{b}\right)^u = \frac{a^u}{b^u}$$

Theorem
$$(a^x)' = a^x \ln a$$

Proof: By Chain Rule, since $\ln a$ is a constant,

$$(a^x)' = (e^{\ln a^x})' = e^{\ln a^x}(x \ln a)' = a^x \ln a$$

Example 1

Find $f'(x)$ when $f(x) = (x^2 + 2)^{10} + 10^{x^2+2}$

$$f'(x) = 10(x^2 + 2)^9 (2x) + (10^{x^2+2} \ln 10)(2x)$$

Theorem
$$\int a^x dx = \left(\frac{1}{\ln a}\right) a^x + C$$

Proof

$$\left(\left(\frac{1}{\ln a}\right) a^x\right)' = \left(\frac{1}{\ln a}\right) a^x \ln a = a^x$$

Example 2

a) $\int 5^x dx = \left(\frac{1}{\ln 5}\right) 5^x + C$

b) $\int x 5^{x^2} dx = \frac{1}{2} \int 5^u du = \frac{1}{2}\left(\frac{1}{\ln 5}\right) 5^u + C$

$= \left(\frac{1}{2\ln 5}\right) 5^{x^2} + C$

$u = x^2$
$du = 2x\, dx$
$x\, dx = \dfrac{du}{2}$

The Definition of $\log_a x$:

$y = \log_a x$ if and only if $x = a^y$

Change of Base Formula: $\log_a x = \dfrac{\ln x}{\ln a}$

Proof: $y = \log_a x$
$x = a^y$
$\ln x = y \ln a$
$y = \dfrac{\ln x}{\ln a} \Rightarrow \log_a x = \dfrac{\ln x}{\ln a}$

Theorem
$$(\log_a x)' = \left(\frac{\ln x}{\ln a}\right)' = \frac{1}{\ln a} \cdot \frac{1}{x}$$

Example 3

If $f(x) = \log \sqrt[3]{(2x+3)^2}$, find $f'(x)$

$f'(x) = \left(\log(2x+3)^{\frac{2}{3}}\right)' = \left(\frac{2}{3}\log|2x+3|\right)'$

$= \frac{2}{3} \cdot \frac{1}{\ln 10} \cdot \frac{1}{2x+3}(2) = \dfrac{4}{3(2x+3)\ln 10}$

Example 4

If $y = x^x$ and $x > 0$, find $\dfrac{dy}{dx}$

$y = x^x$ — Use Log differentiation

$\ln y = x \ln x$ — Take ln on both sides

$\dfrac{1}{y}\dfrac{dy}{dx} = x(\ln x)' + \ln x \cdot x'$ — Implicitly differentiate

$\dfrac{dy}{dx} = (1 + \ln x) y$ — Multiply both sides by y

$\dfrac{dy}{dx} = (1 + \ln x) \cdot x^x$ — Replace y with original function

Example 5

If $y = x^{3x+4}$ and $x > 0$, find $\dfrac{dy}{dx}$

$y = x^{3x+4}$ — Use Log differentiation

$\ln y = \ln x^{3x+4}$ — Take ln on both sides

$(\ln y)' = ((3x+4)\ln x)'$ — Implicitly differentiate

$\dfrac{1}{y} \cdot \dfrac{dy}{dx} = (3x+4)' \ln x + (\ln x)'(3x+4)$ — Use Product Rule on the right and Chain Rule on the left

$\dfrac{1}{y} \cdot \dfrac{dy}{dx} = 3\ln x + \dfrac{1}{x}(3x+4)$ — Find each derivative

$\dfrac{dy}{dx} = \left(3\ln x + 3 + \dfrac{4}{x}\right) \cdot x^{3x+4}$ — Multiply both sides by y; Replace y with original function

Example 6

Find the volume generated when the region bounded by the curves $y = 2^x$ and $y = x^2$ and lines $x = 0$ and $x = 2$ is rotated about the x-axis.

1) Find the points of intersection
$2^x = x^2 \Rightarrow x = 2$

2) Set up the integral on $[0,2]$

Washers Method: $V = \pi \int_a^b (R^2 - r^2)\,dx$

$V = \pi \int_0^2 \left[(2^x)^2 - (x^2)^2\right] dx = \pi \int_0^2 \left[(4^x) - (x^4)\right] dx$

3) Evaluate the integral

$= \pi \left[\dfrac{4^x}{2\ln 2} - \dfrac{x^5}{5}\right]_0^2 = \pi \left(\dfrac{16}{2\ln 2} - \dfrac{32}{5} - \dfrac{1}{2\ln 2}\right)$

$= \boxed{\pi \left(\dfrac{15}{2\ln 2} - \dfrac{32}{5}\right)}$

Theorem

$\boxed{(I)\ \lim_{h \to 0}(1+h)^{\frac{1}{h}} = e}$ or, if $n = \dfrac{1}{h}$: $(II)\ \lim_{n \to \infty}\left(1 + \dfrac{1}{n}\right)^n = e$

Proof: $f(x) = \ln x$

Use limit definition of derivative:

$f'(x) = \lim_{h \to 0} \dfrac{\ln(x+h) - \ln x}{h}$

$= \lim_{h \to 0} \dfrac{1}{h} \ln \dfrac{x+h}{x}$ — Use property of logarithms

$= \lim_{h \to 0} \dfrac{1}{h} \ln\left(1 + \dfrac{h}{x}\right)$ — Simplify expression inside of ln

Proof of the Theorem continued

When $x = 1$,

$f'(x) = \lim_{h \to 0} \ln(1+h)^{\frac{1}{h}}$

$f'(x) = \dfrac{1}{x} \Rightarrow f'(1) = 1$

$\lim_{h \to 0} \ln(1+h)^{\frac{1}{h}} = 1$

Switch places lim and ln

$\ln \lim_{h \to 0}(1+h)^{\frac{1}{h}} = 1 \quad \Rightarrow \quad \boxed{(I)\ \lim_{h \to 0}(1+h)^{\frac{1}{h}} = e^1}$

Use this space for notes.

6.6 Laws of Growth and Decay

 Mathboat.com

Theorem

Let y be a differentiable function such that $y > 0$ for every t, y_0 is the value of y at $t = 0$.

If $\dfrac{dy}{dt} = cy$ for some constant c, then: $y = y_0 e^{ct}$

Proof

$\dfrac{dy}{dt} = cy$ \qquad $y = B \cdot e^{ct}$

$\displaystyle\int \dfrac{dy}{y} = \int c\,dt$ \quad Separate the variables and integrate \qquad $y_0 = Be^{c(0)} = B$

$\ln y = ct + b$ \qquad $\boxed{y = y_0 e^{ct}}$

$y = e^{ct+b} = e^{ct} \cdot e^{b}$

Example 1

The number of bacteria in a culture increases from 400 to 1600 in three hours. Assuming that the rate of increase is directly proportional to the number of bacteria present, find:

a. a formula for the number of bacteria at time t

$\dfrac{dy}{dt} = cy \Rightarrow y_t = y_0 e^{ct}$

$y = 400 e^{ct}$

$1600 = 400 e^{3c}$

$e^{3c} = 4 \Rightarrow e^c = 4^{\frac{1}{3}}$

$y = 400 e^{ct} \Rightarrow \boxed{y = 400(4)^{\frac{t}{3}}}$

b. the number of bacteria at $t = 6$

$y = 400(4)^{\frac{t}{3}} = 400(4)^{\frac{6}{3}} = 400(16) = \boxed{6400}$

Given:
$y(0) = 400$
$y(3) = 1600$
$\dfrac{dy}{dt} = cy$
$y(t) = ?$
$y(6) = ?$

Example 2

Radioactive substance decays exponentially and has a half-life of approximately 2000 years.

a. What is the amount y remaining from 40 milligrams of this substance after t years?

$y = 40 e^{ct}$

$20 = 40 e^{2000c}$

$e^{2000c} = \dfrac{1}{2} \Rightarrow e^c = 2^{\frac{-1}{2000}}$

plug into $y = 40 e^{ct} \Rightarrow y = 40(2)^{\frac{-t}{2000}}$

b. when will the amount remaining be 10mg?

$10 = 40(2)^{\frac{-t}{2000}}$

$2^{-\frac{t}{2000}} = \dfrac{10}{40} \Rightarrow 2^{\frac{t}{2000}} = 4$, take ln on both sides

$\Rightarrow \dfrac{t}{2000} \ln 2 = \ln 4 \Rightarrow t = 4000\,yr$

Given:
$y = y_0 e^{ct}$
$y(0) = 40$
$y(2000) = \dfrac{1}{2}(40) = 20$
$y(t) = ?$
$y(?) = 20$

Example 3.
The maximum population of the earth is 50 Billion people. In 1980 the population was 4.5 billion. Assuming that the population increases at a rate of 2% and the rate of increase is directly proportional to the number of people, in how many years will the maximum population be reached?

$\dfrac{dP}{dt} = .02P$

$\displaystyle\int \dfrac{dP}{P} = \int .02\,dt$

$\ln|P| = .02t + k$

$P = e^{.02t + k} = e^{.02t} e^{k}$

$P_0 = e^0 e^k = e^k$

$P = P_0 e^{.02t}$

$\boxed{P = 4.5 e^{.02t}}$

$50 = 4.5 e^{.02t}$

$\dfrac{50}{4.5} = e^{.02t}$

$\ln\left(\dfrac{50}{4.5}\right) = \ln e^{.02t}$

$\ln\left(\dfrac{50}{4.5}\right) = .02t$

$t = \dfrac{\ln\left(\dfrac{50}{4.5}\right)}{.02} \Rightarrow \boxed{t \approx 120.397 \text{ yrs.}}$

Given:
$P_0 = 4.5$
$c = 0.02$
$\dfrac{dP}{dt} = cP$
$P(?) = 50$

Example 4.
Newton's law of cooling states that the rate at which an object cools is directly proportional to the difference in temperature between the object and the surrounding medium. If a pie cools from 120°F to 90°F in half an hour with surrounding temperature at 70°F, what is the pie's temperature at the end of the next half hour?

$\dfrac{dy}{dt} = c(y - 70)$

$\displaystyle\int \dfrac{1}{(y-70)} dy = \int c\,dt$

$\ln(y-70) = ct + b$

$y - 70 = e^{ct+b} = e^b e^{ct}$

$120 - 70 = e^b e^0 = e^b \Rightarrow e^b = 50$

Plug into $y - 70 = e^b e^{ct}$

$y - 70 = 50 e^{ct}$ or $y = 50 e^{ct} + 70$

$90 = 50 e^{\frac{c}{2}} + 70$

$e^{\frac{c}{2}} = \dfrac{20}{50} = \dfrac{2}{5} \Rightarrow e^c = \dfrac{4}{25}$

Let $y =$ temperature of the pie after t hours of cooling. Temperature of the surrounding medium is 70°, so the difference in temperature is $y - 70$.

Substitute into
$y = 50 e^{ct} + 70$

$y = 50\left(\dfrac{4}{25}\right)^t + 70$

$y(1) = 50\left(\dfrac{4}{25}\right)^1 + 70$

$= \boxed{78°F}$

Given:
$\dfrac{dy}{dt} = c(y - 75)$
$y(0) = 120$
$y\left(\dfrac{1}{2}\right) = 90$
$y\left(\dfrac{1}{2} + \dfrac{1}{2}\right) = y(1) = ?$

6.7 Logistic Population Growth

Mathboat.com

Logistic population growth

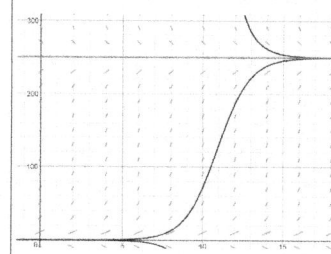

The **logistic model** is a method to model population change, taking into account the **presence of limited resources or space**.

Letting P represent population size (N is often used in ecology instead) and t represent time, this model is formalized by the differential equation:

$$\frac{dP}{dt} = rP\left(1 - \frac{P}{K}\right)$$

where the constant r defines the growth rate and K is the **Carrying Capacity**, the maximum number of individuals in a population that the environment can support.

$$\frac{dP}{dt} = rP\left(1 - \frac{P}{K}\right)$$

If P is small compared to K,

then the term $1 - \frac{P}{K}$ is close to 1, so $\frac{dP}{dt} \approx rP$
(like exponential growth).

As P approaches K, then the term $1 - \frac{P}{K}$ approaches 0,

so $\frac{dP}{dt}$ approaches 0.

(growth slows down as the population approaches the carrying capacity).

Logistic population growth

$$\frac{dP}{dt} = rP\left(1 - \frac{P}{K}\right)$$

Notice that if the population P is smaller than carrying capacity K,

$$P < K \Rightarrow 1 - \frac{P}{K} > 0 \Rightarrow \frac{dP}{dt} > 0$$

Then, the population P will grow if it starts below its carrying capacity.

By an analogous argument, the population will decrease (and have a negative rate of growth) if it starts out higher than K.

Example 1
This is not a Logistic population growth, because of t in the equation, not P

The population $P(t)$ of a species satisfies the logistic differential

equation $\frac{dP}{dt} = P\left(5 - \frac{t}{10}\right)$. Find $P(t)$ if $P(0) = 20$.

$\frac{dP}{dt} = P\left(5 - \frac{t}{10}\right)$ Separate the variables:

$\int \frac{1}{P} dP = \int \left(5 - \frac{t}{10}\right) dt \Rightarrow \ln|P| = 5t - \frac{t^2}{20} + C$

Plug in Initial Conditions

$\ln|20| = 0 - 0 + C \Rightarrow C = \ln 20$

$\ln|P| = 5t - \frac{t^2}{20} + \ln 20$

$P = e^{5t - \frac{t^2}{20} + \ln 20} = e^{5t - \frac{t^2}{20}} e^{\ln 20}$ $\boxed{P = 20e^{5t - \frac{t^2}{20}}}$

Example 2.
The population $P(t)$ of a species satisfies the logistic

differential equation $\frac{dP}{dt} = P\left(5 - \frac{P}{10}\right)$ where the initial population

$P(0) = 20$ and t is the time in years. What is $\lim\limits_{t \to \infty} P(t)$?

This is the Logistic population growth!

$$\frac{dP}{dt} = P\left(5 - \frac{P}{10}\right)$$

$$\frac{dP}{dt} = \frac{P(50 - P)}{10}$$

$$\int \frac{10 \, dP}{P(50 - P)} = \int dt$$

$$10 \int \frac{dP}{P(50 - P)} = \int dt$$

Example 2 Continued

Using Partial Fractions,
(read more in section 8.3),

$$\frac{1}{P(50-P)} = \frac{A}{P} + \frac{B}{50-P}$$

Set $P = 0 : A = \frac{1}{50}$

Set $P = 50 : B = \frac{1}{50}$

$$10\int \frac{dP}{P(50-P)} = 10\int \frac{1}{50P} + \frac{1}{50(50-P)} dP = \int dt$$

$$t = \frac{1}{5}\ln|P| - \frac{1}{5}\ln|50-P| + C \Rightarrow \boxed{t = \frac{1}{5}\ln\left|\frac{P}{50-P}\right| + C}$$

Example 2 Continued

Plug in Initial Conditions, $P(0) = 20$

$$0 = \frac{1}{5}\ln\left|\frac{20}{30}\right| + C \Rightarrow C = -\frac{1}{5}\ln\left|\frac{2}{3}\right| = \ln\left(\frac{3}{2}\right)^{\frac{1}{5}}$$

$$t = \frac{1}{5}\ln\left|\frac{P}{50-P}\right| + \ln\left(\frac{3}{2}\right)^{\frac{1}{5}}$$

$$\ln\left|\frac{P}{50-P}\right|^{\frac{1}{5}} = t - \ln\left(\frac{3}{2}\right)^{\frac{1}{5}}$$

$$\left|\frac{P}{50-P}\right|^{\frac{1}{5}} = e^{t-\ln\left(\frac{3}{2}\right)^{\frac{1}{5}}}$$

$$\left|\frac{P}{50-P}\right| = e^{5\left(t-\ln\left(\frac{3}{2}\right)^{\frac{1}{5}}\right)}$$

Example 2 Continued

As $t \to \infty$

$$e^{5\left(t-\ln\left(\frac{3}{2}\right)^{\frac{1}{5}}\right)} = e^{\infty} = \infty$$

$$\frac{P}{50-P} = \infty$$

$$50 - P = 0 \Rightarrow \boxed{P = 50}$$

It is too long...

But There is an Easier Way!!!

Given: $\frac{dP}{dt} = P\left(5 - \frac{P}{10}\right)$. We want to know $\lim_{t\to\infty} P(t)$.

Because $P(t)$ converges to a certain value as t approaches infinity, we know that the rate of change of the population approaches zero

$$\frac{dP}{dt} = P\left(5 - \frac{P}{10}\right) = 0$$

$P = 0$ or $5 - \frac{P}{10} = 0$

$$\boxed{P = 50}$$

It is much easier!

Example 3.
The population $P(t)$ of a species satisfies the logistic differential equation $\frac{dP}{dt} = P\left(5 - \frac{P}{10}\right)$. If $P(0) = 20$, for what value of P is the population growing the fastest?

$\frac{dP}{dt}$ is the rate at which the population is growing. Therefore, the population is growing the fastest when $\frac{dP}{dt}$ has a maximum.

$$\frac{dP}{dt} = P\left(5 - \frac{P}{10}\right) = 5P - \frac{P^2}{10}$$

To maximize $\frac{dP}{dt}$, find it's derivative, $\frac{d^2P}{dt^2}$, and set it $= 0$

Example 3 Continued

$$\frac{d^2P}{dt^2} = 5\frac{dP}{dt} - \frac{1}{5}P\frac{dP}{dt}$$

$$\frac{d^2P}{dt^2} = \frac{dP}{dt}\left(5 - \frac{P}{5}\right)$$

$$5 - \frac{P}{5} = 0 \quad \boxed{P = 25}$$

$\frac{d^2P}{dt^2}$ 	+ max −
          $P = 25$
          local and absolute

P = 25 is the absolute max because the function is increasing and then decreasing

7.1 Inverse Trigonometric Functions

Definition

Inverse sine function is $y = \text{Sin}^{-1} x$ if and only if $x = \sin y$ for $[-1,1]$ as domain and $\left[-\dfrac{\pi}{2}, \dfrac{\pi}{2}\right]$ as range.

Graph of $y = \text{Sin}^{-1} x$ is the image of $y = \sin x$ reflected over $y = x$

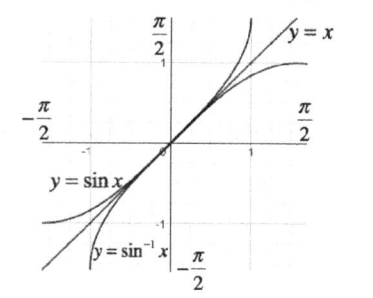
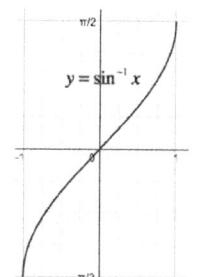

Properties of Arcsin x

1) $\sin(\text{Sin}^{-1} x) = \sin(\text{Arcsin } x) = x$ if $-1 \le x \le 1$
2) $\text{Sin}^{-1}(\sin x) = \text{Arcsin}(\sin x) = x$ if $-\dfrac{\pi}{2} \le x \le \dfrac{\pi}{2}$

x should be inside of the domain of inner function

Examples

1. $\sin\left(\text{Sin}^{-1} \dfrac{1}{3}\right) = \dfrac{1}{3}$, since $-1 < \dfrac{1}{3} < 1$

$\left(\dfrac{1}{3} \text{ is inside of domain of Arcsine}\right)$

Examples Continued

2. $\text{Arcsin}\left(\sin \dfrac{\pi}{3}\right) = \dfrac{\pi}{3}$, since $-\dfrac{\pi}{2} < \dfrac{\pi}{3} < \dfrac{\pi}{2}$

$\left(\dfrac{\pi}{3} \text{ is inside of restricted domain of sine}\right)$

2nd quadrant

3. $\text{Sin}^{-1}\left(\sin \dfrac{3\pi}{4}\right) = \text{Sin}^{-1}\left(\sin \dfrac{\pi}{4}\right) = \dfrac{\pi}{4}$ since $-\dfrac{\pi}{2} < \dfrac{\pi}{4} < \dfrac{\pi}{2}$

$\left(\dfrac{\pi}{4} \text{ is inside of restricted domain of sine}\right)$

Definition

Inverse cosine function is $y = \text{Cos}^{-1} x$ if and only if $x = \cos y$ for $[-1,1]$ as domain and $[0, \pi]$ as range.

Graph of $y = \text{Cos}^{-1} x$ is the image of $y = \cos x$ reflected over $y = x$.

Since $y = \cos x$ has to pass horizontal test in order to have an inverse function, we need to restrict the domain of $y = \cos x$ from 0 to π.

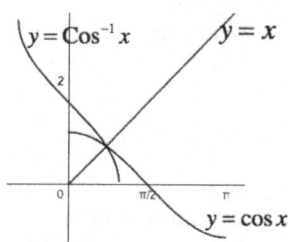
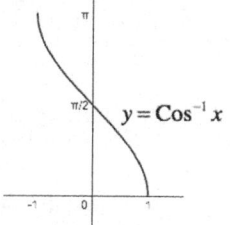

Properties of Arccos x

1) $\cos(\text{Cos}^{-1}x) = \cos(\text{Arccos } x) = x$ if $-1 \le x \le 1$
2) $\text{Cos}^{-1}(\cos x) = \text{Arccos}(\cos x) = x$ if $0 \le x \le \pi$

x should be inside of the domain of inner function

Examples

1) $\cos\left(\text{Cos}^{-1}\left(-\dfrac{1}{3}\right)\right) = -\dfrac{1}{3}$ since $-1 < -\dfrac{1}{3} < 1$

$\left(-\dfrac{1}{3} \text{ is inside of domain of Arccosine}\right)$

Examples Continued

2. $\text{Cos}^{-1}\left(\cos\dfrac{5\pi}{6}\right) = \dfrac{5\pi}{6}$ since $0 < \dfrac{5\pi}{6} < \pi$

$\left(\dfrac{5\pi}{6} \text{ is inside of restricted domain of cosine}\right)$

even function

3. $\text{Cos}^{-1}\left(\cos\left(-\dfrac{\pi}{6}\right)\right) = \text{Cos}^{-1}\left(\cos\dfrac{\pi}{6}\right) = \dfrac{\pi}{6}$ since $0 < \dfrac{\pi}{6} < \pi$

$\left(\dfrac{\pi}{6} \text{ is inside of restricted domain of cosine}\right)$

Definition

The inverse tangent function, or arctangent is
$y = \text{Tan}^{-1}x$ if and only if $x = \tan y$
for every x and $\left(-\dfrac{\pi}{2}, \dfrac{\pi}{2}\right)$ as range

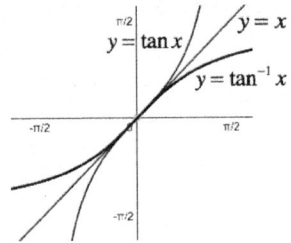
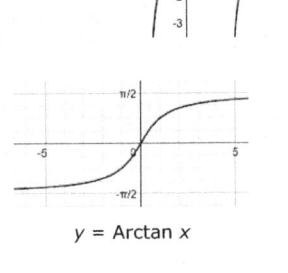

$y = \text{Arctan } x$

Properties of Arctan x

$\tan(\text{Tan}^{-1}x) = \tan(\text{Arctan } x) = x$ for every x

$\text{Tan}^{-1}(\tan x) = \text{Arctan}(\tan x) = x$ if $-\dfrac{\pi}{2} < x < \dfrac{\pi}{2}$

x should be inside of the domain of inner function

Examples

1) $\tan\left(\text{Tan}^{-1}\dfrac{1}{2}\right) = \tan\left(\text{Arctan}\dfrac{1}{2}\right) = \dfrac{1}{2}$

Examples of Properties of Arctan x (continued)

2) $\text{Tan}^{-1}\left(\tan\dfrac{\pi}{3}\right) = \text{Arctan}\left(\tan\dfrac{\pi}{3}\right) = \dfrac{\pi}{3}$

3) $\text{Tan}^{-1}\left(\tan\dfrac{5\pi}{6}\right) = \text{Tan}^{-1}\left(-\tan\dfrac{\pi}{6}\right) =$

$\text{Tan}^{-1}\left(\tan\left(-\dfrac{\pi}{6}\right)\right) = -\dfrac{\pi}{6}$

$\left(\dfrac{5\pi}{6} \text{ is not inside of restricted domain of tangent}\right)$

Example 1

Find the exact value of $\sec\left(\text{Arctan}\dfrac{2}{5}\right)$.

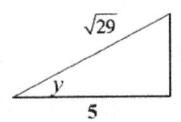

Solution

This is the question we need to answer:

What is secant of an angle, which has tangent $= \dfrac{2}{5}$?

Let $y = \text{Arctan}\dfrac{2}{5}$, then $\tan y = \dfrac{2}{5}$ $\sec y = ?$

By Pythagorean theorem: hypotenuse $= \sqrt{5^2 + 2^2} = \sqrt{29}$

$\boxed{\sec\left(\text{Arctan}\dfrac{2}{5}\right) = \sec y = \dfrac{\sqrt{29}}{5}}$

Example 2.
Find the exact value of $\cos\left(\tan^{-1}\left(-\frac{2}{3}\right)\right)$.

This is the question we need to answer:

What is cos of the angle which has a tangent $=-\frac{2}{3}$?

So, $\tan x = -\frac{2}{3}$; Tan is negative in quadrants II and IV.

Restrictions for domain of $\tan x$ are: $\left(-\frac{\pi}{2}, \frac{\pi}{2}\right)$ or I and IV.

So, it is Quadrant IV.

$\cos x$ is positive in IV.

$$\boxed{\cos x = \frac{3}{\sqrt{13}} = \frac{3\sqrt{13}}{13}}$$

Example 3
Find the exact value of $\tan\left(\text{Arcsin}\,-\frac{\sqrt{3}}{2}\right)$.

This is the question we need to answer:

What is tangent of the angle which has a sine $=-\frac{\sqrt{3}}{2}$?

sin is negative in III, IV; Restrictions: $\left[-\frac{\pi}{2}, \frac{\pi}{2}\right]$

So, it is quadrant IV; $\tan\theta = -\frac{\sqrt{3}}{1} = \boxed{-\sqrt{3}}$

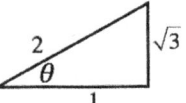

Example 4
Find the exact value of

$$\sin\left(\text{Arctan}\left(-\frac{1}{2}\right) - \text{Arccos}\left(\frac{4}{5}\right)\right).$$

Let $u = \text{Arctan}(-1/2)$ and $v = \text{Arccos}\,4/5$
$\tan u = -1/2$ and $\cos v = 4/5$
Tangent is negative in II and IV

Restrictions for its domain is $\left(-\frac{\pi}{2}, \frac{\pi}{2}\right)$ or: I and IV,

so it is quadrant IV.

Example 4 Continued

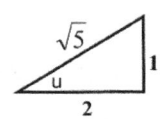 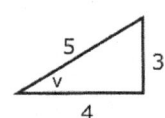

$\sin u = -\frac{1}{\sqrt{5}}$ $\sin v = \frac{3}{5}$

$\cos u = +\frac{2}{\sqrt{5}}$ $\cos v = \frac{4}{5}$

$\sin(u-v) = \sin u \cos v - \cos u \sin v$

$= -\frac{1}{\sqrt{5}}\left(\frac{4}{5}\right) - \frac{2}{\sqrt{5}}\left(\frac{3}{5}\right) = \frac{-4-6}{5\sqrt{5}} = \frac{-2}{\sqrt{5}} = \boxed{\frac{-2\sqrt{5}}{5}}$

Example 5
If $-1 \le x \le 1$, rewrite $\cos(\text{Sin}^{-1}x)$ as an algebraic expression in x.

Solution

Let $y = \text{Sin}^{-1}x$ or, $\sin y = x$

$\cos y = \sqrt{1-\sin^2 y} = \sqrt{1-x^2}$

then: $\cos(\text{Sin}^{-1}x) = \sqrt{1-x^2}$

OR draw the triangle for $\sin y = x$:

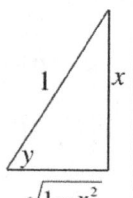

$\cos(\text{Sin}^{-1}x) = \cos y = \frac{\sqrt{1-x^2}}{1} = \boxed{\sqrt{1-x^2}}$

Example 6
Find solutions of $3\sin^2 t + 5\sin t - 2 = 0$ on the interval $[-\pi/2, \pi/2]$

Solution: Equation is quadratic with $x = \sin t$

$$3x^2 + 5x - 2 = 0$$

$x = \frac{-b \pm \sqrt{b^2-4ac}}{2a} \Rightarrow \sin t = \frac{-5 \pm \sqrt{25+24}}{6} = \frac{-5 \pm 7}{6} = \cancel{-2} \text{ and } \frac{1}{3}$

$\boxed{t = \text{Sin}^{-1}\frac{1}{3} \approx 0.3398} \in [-\pi/2, \pi/2]$

7.2 Derivatives of Inverse Trigonometric Functions.

Mathboat.com

Derivative of Arcsin x

$(\sin^{-1} x)' = \dfrac{1}{\sqrt{1-x^2}}$ Proof

$y = \sin^{-1} x \quad -1 < x < 1, \quad -\dfrac{\pi}{2} < y < \dfrac{\pi}{2}$

$\sin y = x$ Implicitly differentiate

$\cos y \cdot \dfrac{dy}{dx} = 1$

$\dfrac{dy}{dx} = \dfrac{1}{\cos y} = \boxed{\dfrac{1}{\sqrt{1-x^2}}}$

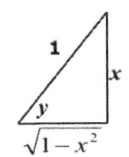

Derivative of Arctan x

$\boxed{(\tan^{-1} x)' = \dfrac{1}{1+x^2}}$ Proof

$y = \tan^{-1} x, \quad -\dfrac{\pi}{2} < y < \dfrac{\pi}{2}$

$\tan y = x$

Implicitly differentiate

$\sec^2 y \cdot \dfrac{dy}{dx} = 1$

$\dfrac{dy}{dx} = \dfrac{1}{\sec^2 y} = \cos^2 y = \dfrac{1}{1+x^2}$

Derivative of Arcsec x

$\boxed{(\sec^{-1} x)' = \dfrac{1}{x\sqrt{x^2-1}}}$ Proof

$y = \sec^{-1} x$

$\sec y = x$ for y in either $\left(0, \dfrac{\pi}{2}\right)$ or $\left(\pi, \dfrac{3\pi}{2}\right)$.

Implicitly differentiate.

$\sec y \tan y \cdot \dfrac{dy}{dx} = 1$

$\Rightarrow \dfrac{dy}{dx} = \dfrac{1}{\sec y \tan y}$

$= \dfrac{1}{x\sqrt{x^2-1}}$ for $|x| > 1$

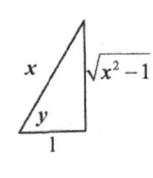

Examples on Derivatives of the Inverse Trig Functions

$\boxed{\begin{array}{ll}(\sin^{-1} u)' = \dfrac{1}{\sqrt{1-u^2}} \cdot u' & (\tan^{-1} u)' = \dfrac{1}{1+u^2} \cdot u' \\ (\cos^{-1} u)' = -\dfrac{1}{\sqrt{1-u^2}} \cdot u' & (\sec^{-1} u)' = \dfrac{1}{u\sqrt{u^2-1}} \cdot u'\end{array}}$

Example 1

$(\sin^{-1} 5x)' = \dfrac{1}{\sqrt{1-(5x)^2}} \cdot (5x)' = \dfrac{5}{\sqrt{1-25x^2}}$

Example 2

$[\arccos(\tan x)]' = -\dfrac{1}{\sqrt{1-(\tan x)^2}} (\tan x)' = -\dfrac{\sec^2 x}{\sqrt{1-\tan^2 x}}$

Example 3

$(\tan^{-1}(5x^2))' = \dfrac{1}{1+(5x^2)^2} (5x^2)' = \dfrac{10x}{1+25x^4}$

Example 4

$[\arcsec(x^5)]' = \dfrac{1}{x^5\sqrt{(x^5)^2-1}} (x^5)' = \dfrac{5x^4}{x^5\sqrt{x^{10}-1}} = \dfrac{5}{x\sqrt{x^{10}-1}}$

7.3 Integrals Involving Inverse Trigonometric Functions.

Derivatives of Inverse Trigonometric Functions

$$\left(\sin^{-1} u\right)' = \frac{1}{\sqrt{1-u^2}} \cdot u'$$

$$\left(\sec^{-1} u\right)' = \frac{1}{u\sqrt{u^2-1}} \cdot u'$$

$$\left(\tan^{-1} u\right)' = \frac{1}{1+u^2} \cdot u'$$

Then Integration formulas formulas for $a > 0$ will be:

Integrals involving Inverse Trigonometric Functions

$$\int \frac{1}{\sqrt{a^2-u^2}} du = \sin^{-1}\frac{u}{a} + C$$

$$\int \frac{1}{u\sqrt{u^2-a^2}} du = \frac{1}{a}\sec^{-1}\frac{u}{a} + C$$

$$\int \frac{1}{a^2+u^2} du = \frac{1}{a}\tan^{-1}\frac{u}{a} + C$$

Proof of one of the integration formulas

Prove: $\int \frac{1}{\sqrt{a^2-u^2}} du = \sin^{-1}\frac{u}{a} + C$

$\int \frac{1}{\sqrt{a^2-u^2}} du = \int \frac{1}{\sqrt{a^2\left(1-\frac{u^2}{a^2}\right)}} du$

$= \int \frac{1}{a\sqrt{1-\frac{u^2}{a^2}}} du = \int \frac{1}{\sqrt{1-y^2}} dy$

$= \sin^{-1} y + C = \boxed{\sin^{-1}\frac{u}{a} + C}$

$\boxed{\dfrac{u}{a} = y \\ \dfrac{1}{a} du = dy}$

$\left(\sin^{-1} y\right)' = \dfrac{1}{\sqrt{1-y^2}}$

Example 1

Evaluate $\int \dfrac{4e^{3x}}{\sqrt{1-e^{6x}}} dx$

$= \dfrac{4}{3} \int \dfrac{1}{\sqrt{1-u^2}} du = \dfrac{4}{3}\sin^{-1} u + C$

$= \boxed{\dfrac{4}{3}\sin^{-1} e^{3x} + C}$

$\boxed{u = e^{3x} \\ du = 3e^{3x} dx \\ e^{3x} dx = \dfrac{du}{3}}$

$\int \dfrac{1}{\sqrt{a^2-u^2}} du = \sin^{-1}\dfrac{u}{a} + a$

$a^2 = 1$

Example 2

Evaluate $\int \dfrac{2x^3}{7+x^8} dx$

$\dfrac{2}{4} \int \dfrac{1}{\left(\sqrt{7}\right)^2 + u^2} du$

$= \left(\dfrac{1}{2}\right)\dfrac{1}{\sqrt{7}}\tan^{-1}\dfrac{u}{\sqrt{7}} + C$

$= \boxed{\dfrac{\sqrt{7}}{14}\tan^{-1}\dfrac{x^4}{\sqrt{7}} + C}$

$\boxed{u = x^4 \\ du = 4x^3 dx \\ x^3 dx = \dfrac{du}{4}}$

$\int \dfrac{1}{a^2+u^2} du = \dfrac{1}{a}\tan^{-1}\dfrac{u}{a} + C$

$a^2 = 7$

Example 3

Evaluate $\int \dfrac{1}{x\sqrt{x^6-16}} dx$

$= \int \dfrac{1 \cdot x^2}{\boxed{x^2 \cdot x} \cdot \sqrt{\left(x^3\right)^2 - 4^2}} dx$

$= \dfrac{1}{3} \int \dfrac{1}{u\sqrt{u^2 - 4^2}} du$

$= \dfrac{1}{3} \cdot \dfrac{1}{4}\sec^{-1}\dfrac{u}{4} + C$

$= \boxed{\dfrac{1}{12}\sec^{-1}\dfrac{x^3}{4} + C}$

$\boxed{u = x^3 \\ du = 3x^2 dx \\ x^2 dx = \dfrac{du}{3}}$

We need to have $\boxed{x^3}$ in front of the square root, so multiply numerator and denominator by x^2

$\int \dfrac{1}{u\sqrt{u^2-a^2}} du = \dfrac{1}{a}\sec^{-1}\dfrac{u}{a} + C$

$a^2 = 16$

Formulas and Theorems

Limits

$\lim_{x \to c}[f(x)+g(x)] = \lim_{x \to c} f(x) + \lim_{x \to c} g(x)$	$\lim_{x \to c}[af(x)] = a \lim_{x \to c} f(x)$
$\lim_{x \to c}[f(x) \cdot g(x)] = \lim_{x \to c} f(x) \cdot \lim_{x \to c} g(x)$	$\lim_{x \to c}\left[f(x)^{\frac{a}{b}}\right] = \left[\lim_{x \to c} f(x)\right]^{\frac{a}{b}}$
$\lim_{x \to c}\left[\dfrac{f(x)}{g(x)}\right] = \dfrac{\lim_{x \to c} f(x)}{\lim_{x \to c} g(x)}$	$\lim_{n \to 0}(1+n)^{\frac{1}{n}} = \lim_{n \to \infty}\left(1+\dfrac{1}{n}\right)^n = e$
$\lim_{x \to 0} \dfrac{\sin x}{x} = \lim_{x \to 0} \dfrac{x}{\sin x} = 1$	$\lim_{x \to 0} \dfrac{1-\cos x}{x} = 0$

Differentiation formulas

Power Rule: $(x^n)' = nx^{n-1}$	Product Rule: $(f(x) \cdot g(x))' = f'(x) \cdot g(x) + g'(x) \cdot f(x)$
Reciprocal Rule: $\left(\dfrac{1}{g(x)}\right)' = -\dfrac{g'(x)}{g^2(x)}$	Quotient Rule: $\left(\dfrac{f(x)}{g(x)}\right)' = \dfrac{f'(x) \cdot g(x) - g'(x) \cdot f(x)}{g^2(x)}$
$(\sin x)' = \cos x$	Chain Rule: $(f(g(x)))' = f'(g(x)) \cdot g'(x)$
$(\cos x)' = -\sin x$	$(a)' = 0$
$(\tan x)' = \sec^2 x$	$(x)' = 1$
$(\cot x)' = -\csc^2 x$	$(e^x)' = e^x$
$(\sec x)' = \sec x \tan x$	$(\ln x)' = \dfrac{1}{x}$
$(\csc x)' = -\csc x \cot x$	$(a^x)' = a^x \ln a,\ a > 0,\ a \neq 1$
$(\sin^{-1} x)' = \dfrac{1}{\sqrt{1-x^2}}$	$(\log_a x)' = \dfrac{1}{x \ln a}$
$(\tan^{-1} x)' = \dfrac{1}{1+x^2}$	$(\sqrt{x})' = \dfrac{1}{2\sqrt{x}}$
$(\sec^{-1} x)' = \dfrac{1}{x\sqrt{x^2-1}}$	$\left(\dfrac{1}{x}\right)' = -\dfrac{1}{x^2}$

Formulas and Theorems

Integration Formulas

$\int \sin x \, dx = -\cos x + C$	$\int x^n \, dx = \dfrac{x^{n+1}}{n+1} + C, \; n \neq -1$
$\int \cos x \, dx = \sin x + C$	$\int a \, dx = ax + C$
$\int \tan x \, dx = -\ln|\cos x| + C$ or $\ln|\sec x| + C$	$\int e^x \, dx = e^x + C$
$\int \cot x \, dx = \ln|\sin x| + C$ or $-\ln|\csc x| + C$	$\int a^x \, dx = \dfrac{a^x}{\ln a} + C$
$\int \sec x \, dx = \ln|\sec x + \tan x| + C$	$\int \dfrac{1}{x} \, dx = \ln|x| + C$
$\int \csc x \, dx = \ln|\csc x - \cot x| + C$	$\int \dfrac{1}{\sqrt{a^2 - x^2}} \, dx = \sin^{-1} \dfrac{x}{a} + C$
$\int \sec x \tan x \, dx = \sec x + C$	$\int \dfrac{1}{a^2 + x^2} \, dx = \dfrac{1}{a} \tan^{-1} \dfrac{x}{a} + C$
$\int \csc x \cot x \, dx = -\csc x + C$	$\int \dfrac{1}{x\sqrt{x^2 - a^2}} \, dx = \dfrac{1}{a} \sec^{-1} \dfrac{x}{a} + C$
$\int \sec^2 x \, dx = \tan x + C$	$\int \csc^2 x \, dx = -\cot x + C$

Continuity	A function $f(x)$ is continuous at $x = a$ if all of the following are true: I. $f(a)$ exists II. $\lim\limits_{x \to a} f(x)$ exists III. $\lim\limits_{x \to a} f(x) = f(a)$
Intermediate Value Theorem	If function $f(x)$ is continuous on a closed interval $[a,b]$ and if w is any number between $f(a)$ and $f(b)$, then there is at least one number c in $[a,b]$ such that $f(c) = w$
Limit Theorem	$\lim\limits_{x \to a} f(x) = L$ if and only if $\lim\limits_{x \to a^+} f(x) = L = \lim\limits_{x \to a^-} f(x)$

Formulas and Theorems

Vertical Asymptote	A line $x = a$ is a vertical asymptote of the graph of a function $y = f(x)$ if $\lim_{x \to a^+} f(x) = \pm\infty$ or $\lim_{x \to a^-} f(x) = \pm\infty$
Horizontal Asymptote	A line $y = L$ is a horizontal asymptote of the graph of a function $f(x)$ if $\lim_{x \to \infty} f(x) = L$ or $\lim_{x \to -\infty} f(x) = L$
Average Rate of Change	Average rate of change of $y = f(x)$ on $[a, a+h]$ is $\dfrac{f(a+h) - f(a)}{h}$
Instantaneous Rate of Change	Instantaneous rate of change of $y = f(x)$ on $[a, a+h]$ is: $\lim_{h \to 0}(\text{Average rate of change}) = \lim_{h \to 0} \dfrac{f(a+h) - f(a)}{h}$
Sandwich Theorem	If $f(x) \leq g(x) \leq h(x)$ for all $x \neq c$ in some interval about c, and $\lim_{x \to c} f(x) = \lim_{x \to c} h(x) = L$, then $\lim_{x \to c} g(x) = L$
Definition of Derivative	$f'(x) = \lim_{h \to 0} \dfrac{f(x+h) - f(x)}{h}$ or $f'(x) = \lim_{x \to c} \dfrac{f(x) - f(c)}{x - c}$
Theorem	If a function $f(x)$ is differentiable at $x = a$, then it is continuous at $x = a$
Differentiability on a closed interval	A function f is differentiable on a closed interval $[a,b]$ if f is differentiable on the open interval (a,b) and if both Right-Hand Derivative at a and Left-Hand Derivative at b exist
Rolle's Theorem	Suppose that $f(x)$ is continuous on the closed interval $[a,b]$ and differentiable on the open interval (a,b). If $f(a) = f(b)$, then there is at least one number c between a and b such that $f'(c) = 0$
Mean Value Theorem	If $f(x)$ is continuous on the closed interval $[a,b]$ and differentiable on the open interval (a,b), then there is at least one number c between a and b such that $f'(c) = \dfrac{f(b) - f(a)}{b - a}$
Derivative of the Inverse function	$g'(c) = \dfrac{1}{f'(g(c))}$
Linear approximation of f(x) near x = x₀	$y = f(x_0) + f'(x_0)(x - x_0)$

Formulas and Theorems

Extrema and Concavity

Critical Numbers (Candidates for Extrema)	A number c in the domain of a function f is a critical number of f if either $f'(c)=0$ or $f'(c)$ does not exist
Increasing/Decreasing Functions	If $f(x)$ is differentiable on (a,b) and continuous on $[a,b]$: $f'(x)>0$ on $(a,b) \Rightarrow f(x)$ is increasing on $[a,b]$ $f'(x)<0$ on $(a,b) \Rightarrow f(x)$ is deccreasing on $[a,b]$
Local Minimum	$f(c)$ is the local minimum value of f if $f(c) \leq f(x)$ for every x in an interval I around c
Local Maximum	$f(c)$ is the local maximum value of f if $f(c) \geq f(x)$ for every x in an interval I around c
Absolute Minimum	$f(c)$ is the absolute minimum value of f if $f(c) \leq f(x)$ for every x in the domain of f
Absolute Maximum	$f(c)$ is the absolute maximum value of f if $f(c) \geq f(x)$ for every x in the domain of f
Finding the Absolute Extrema on [a,b]	1. Find Critical numbers. 2. Calculate the function values at the endpoints of $[a,b]$ and at the critical numbers. 3. The largest of these function values is the absolute maximum. The smallest function value is the absolute minimum.
Finding the Local Extrema (First Derivative Test)	Find critical numbers $x = c$ (where $f'(c)=0$ or DNE) Local minimum occurs at $x=c$, where $f'(x)$ changes from negative to positive. Local maximum occurs at $x=c$, where $f'(x)$ changes from positive to negative.
Concavity	If $f''(x)$ exists on (a,b), then: $f''(x)>0 \Rightarrow f(x)$ concave upward in (a,b) $f''(x)<0 \Rightarrow f(x)$ concave downward in (a,b)
Points of Inflection	Points of Inflection of $f(x)$ are the points on the Domain of $f(x)$ where $f''(x)=0$ or DNE and $f''(x)$ changes it's sign passing through them.
Extreme Value Theorem	If f is continuous on a closed interval $[a,b]$, then f has both an absolute minimum and an absolute maximum value on $[a,b]$

Formulas and Theorems

Integrals

Definite Integral. Property 1	$\int_a^b c \cdot f(x)\,dx = c\int_a^b f(x)\,dx$
Definite Integral. Property 2	$\int_a^b f(x)\,dx = -\int_b^a f(x)\,dx$
Definite Integral. Property 3	$\int_a^a f(x)\,dx = 0$
Definite Integral. Property 4	If f is integrable on a closed interval and if a, b and c are any three numbers in the interval, then $\int_a^b f(x)\,dx = \int_a^c f(x)\,dx + \int_c^b f(x)\,dx$
If $f(x) \geq 0$ on $[a,b]$:	If $f(x) \geq 0$ on $[a,b]$ then $\int_a^b f(x)\,dx \geq 0$
If $g(x) \geq f(x)$ on $[a,b]$:	If $g(x) \geq f(x)$ on $[a,b]$ then $\int_a^b f(x)\,dx \geq 0$
If f(x) is an even function:	If f is an Even Function, then $\int_{-a}^{a} f(x)\,dx = 2\int_0^a f(x)\,dx$
If f(x) is an odd function:	If f is an Odd Function, then $\int_{-a}^{a} f(x)\,dx = 0$
Riemann sum	Let $f(x)$ be defined on a closed interval $[a,b]$ and P is any decomposition of $[a,b]$ into subintervals of the form $[x_{k-1}, x_k]$. Riemann sum of $f(x)$ for P is $R_P = \sum_{k=1}^{n} f(w_k)\Delta x_k$ where $w_k \in [x_{k-1}, x_k]$.
Integral as limit of Riemann sums	The definite integral of f from a to b, $\int_a^b f(x)\,dx = \lim_{\|p\| \to 0} \sum_k f(w_k)\Delta x_k$ provided the limit exists where $\|p\|$ is the norm of the partition (largest Δx_k)
Integral as area under the curve	If f is integrable and $f(x) \geq 0$ for every x in $[a,b]$ (where $a < b$), then the area A of the region under the graph of f from a to b is $A = \int_a^b f(x)\,dx$
Fundamental Theorem of Calculus	$\dfrac{d}{dx}\int_a^x f(t)\,dt = f(x)$ and $\int_a^b f'(x)\,dx = f(b) - f(a)$
Velocity	$v(t) = x'(t)$ $v(t) = \int a(t)\,dt$

Formulas and Theorems

Acceleration	$a(t) = v'(t) = x''(t)$
Speed. Increasing and Decreasing speed	$\text{Speed} = |v(t)| = \sqrt{\left(\dfrac{dx}{dt}\right)^2 + \left(\dfrac{dy}{dt}\right)^2}$ $v(t)$ and $a(t)$ have the same signs $\Rightarrow$ speed increases $v(t)$ and $a(t)$ have different signs $\Rightarrow$ speed decreases
Position	$x(t) = \int v(t)\,dt$
Average Velocity	$v_{av} = \dfrac{s(t_2) - s(t_1)}{t_2 - t_1}$
Instantaneous velocity	$v_{inst} = \lim\limits_{t_2 \to t_1} \dfrac{s(t_2) - s(t_1)}{t_2 - t_1}$
Total Distance	$\text{Total Distance} = \int_a^b |v(t)|\,dt$
Net Change or Displacement	$\text{Net Change} = \int_a^b v(t)\,dt$
Area between the curves	$A = \int_a^b (f(x) - g(x))\,dx$, where $f(x) > g(x)$ on $[a,b]$
Volume by Disks	$V = \pi \int_a^b R^2\,dx$ (if region is rotating around horizontal line) or: $V = \pi \int_c^d R^2\,dy$ (if region is rotating around vertical line)
Volume by Washers	$V = \pi \int_a^b (R^2 - r^2)\,dx$ (if region is rotating around horizontal line) or: $V = \pi \int_c^d (R^2 - r^2)\,dy$ (if region is rotating around vertical line)
Volume by Cross Sections	$V = \int_a^b A(x)\,dx$ (if the cross sections are $\perp$ to x-axis) or: $V = \int_c^d A(y)\,dy$ (if the cross sections are $\perp$ to y-axis)
Average value of f	Average Value of $f(x)$ on $[a,b]$ is: $f_{av} = \dfrac{1}{b-a} \int_a^b f(x)\,dx$
L'Hospital's Rule	If $\lim\limits_{x \to c} \dfrac{f(x)}{g(x)}$ is of the form $\dfrac{0}{0}$ or $\dfrac{\infty}{\infty}$, then $\lim\limits_{x \to c} \dfrac{f(x)}{g(x)} = \lim\limits_{x \to c} \dfrac{f'(x)}{g'(x)}$, provided either $\lim\limits_{x \to c} \dfrac{f'(x)}{g'(x)}$ exists or $\lim\limits_{x \to c} \dfrac{f'(x)}{g'(x)} = \infty$
Power-reducing trig formulas	$\sin^2 x = \dfrac{1 - \cos 2x}{2}$ and $\cos^2 x = \dfrac{1 + \cos 2x}{2}$

Made in the USA
Coppell, TX
26 September 2020